The Mathematics of Derivatives

Founded in 1807, John Wiley & Sons is the oldest independent publishing company in the United States. With offices in North America, Europe, Australia, and Asia, Wiley is globally committed to developing and marketing print and electronic products and services for our customers' professional and personal knowledge and understanding.

The Wiley Finance series contains books written specifically for finance and investment professionals as well as sophisticated individual investors and their financial advisors. Book topics range from portfolio management to e-commerce, risk management, financial engineering, valuation, and financial instrument analysis, as well as much more.

For a list of available titles, visit our Web site at www.WileyFinance.com.

The Mathematics of Derivatives

Tools for Designing Numerical Algorithms

ROBERT L. NAVIN

John Wiley & Sons, Inc.

Published by John Wiley & Sons, Inc., Hoboken, New Jersey.
Published simultaneously in Canada.

For general information on our other products and services or for technical support, please contact our Customer Care Department within the United States at (800) 762-2974, outside the United States at (317) 572-3993 or fax (317) 572-4002.

Wiley also publishes its books in a variety of electronic formats. Some content that appears in print may not be available in electronic formats. For more information about Wiley products, visit our Web site at www.wiley.com.

Library of Congress Cataloging-in-Publication Data:
Navin, Robert L.
   The mathematics of derivatives: tools for designing numerical algorithms/Robert L. Navin.
        p. cm. –(Wiley finance series)
   Includes bibliographical references.
   ISBN-13: 978-0-470-04725-5 (cloth)
   ISBN-10: 0-470-04725-9 (cloth)
 1. Derivative securitie–Valuation–Mathematical models. I. Title. II. Series.
   HG6024.A3N37 2007
   332.64'57015181–dc22

                                                        2006017013

10 9 8 7 6 5 4 3 2 1

This book is dedicated with love to Thais Roda Noya.

Contents

Preface

This book is based on a course intended to quickly teach the basics of derivative mathematics to skilled software designers with no knowledge of financial derivatives. I created these notes for my firm—while still a small startup trading-analytics software company—Real Time Risk Systems LLC. We started the business to build a flexible, fast, real-time pricing-and-risk software application for large derivatives trading operations that deal in a variety of products. A significant design feature of the application was to allow complicated models of derivatives to be "plugged-in" without necessarily writing these models into the initial software package.

In the course of developing and using the course from which this book is derived, it became apparent to me that it would be useful to other industry practitioners for a similar purpose: the education and training of programmers (and even trainee quantitative analysts). The professional programmers, of course, have no need of becoming professional "quants." They do, however, need a basic and broad level of understanding of the mathematical formalism of derivatives as quickly as possible. They also need to cover a broad spectrum of material in enough detail to offer a solid grounding and without a lot of mathematical rigor.

I wanted students who took my course—and those who now read this book—to have the ability to talk sensibly to a quantitative analyst and to understand what quants have to say as well. This book is therefore economically designed for the ground it covers in order to save time and focus on the essentials—and without requiring graduate-level mathematics. It is not a rigorous academic text. Nevertheless, all the basics are available, condensed, and in one place. I view this book as more of an engineering textbook than a thorough mathematics treatise.

Acknowledgments

None of the material here is original although most of the formulae have been rederived or written down from memory and are surely stamped with only my own idiosyncrasies. The significant exception is section 8.1.1., which had such a simple and elegant treatment in the original paper by O. Cheyette that there was no simplification that I could supply. There are many excellent textbooks available, including John C. Hull's *Options, Futures and Other Derivatives*, 6th ed., Jonathan E. Ingersoll's *Theory of Financial Decision Making*, and Darrell Duffie's *Dynamic Asset Pricing Theory*. The majority of the material covered here is also covered in those books. Again, the point of coverage of this book is a quick overview for nonquants. For more in-depth and rigorous coverage, I strongly recommend Hull, Ingersoll, Duffie, and many of the other comprehensive derivatives texts.

That said, I acknowledge the great indebtedness that I have to the people from whom I have learned financial derivatives, namely Gunnar Klinkhammer and Barry Ryan at CMS in Los Angeles, who taught me risk-neutral pricing; and Scott Waltz, Albert Sizook, Bill Cherney, Amitabha Sen, and Guillermo Bubliek at Swiss Bank/O'Connor in Chicago from whom Jeff Miller and I learned (and discussed with each other) the elements of finance theory and application, including numerical techniques. The work of Jean-Philippe Bouchaud, which came to me from my discussions with Jeff Miller, also had a significant impact on my understanding of risk-neutral pricing and hedging strategies. I would like to thank Rosie Rush for early copyediting and helping me to put this work together for publication. Many thanks to Pamela Van Giessen at John Wiley and Sons for her help and support for this project.

I would like to thank my colleagues Chris Leon and Nikolai Avteniev for being the first canaries in the mineshaft on whom I tried out the lecture course from which these notes arose. I would also like to thank Rosie Rush for early copyediting and helping me to put this work together for publication. Many thanks to Pamela Van Giessen and Jennifer MacDonald at John Wiley and Sons for their help and support for this project.

The Mathematics of Derivatives

PART I

The Models

1

The Models

CHAPTER 1

Introduction to the Techniques of Derivative Modeling

1.1. INTRODUCTION

"How do I model derivatives?"

For the desk quantitative analyst ("desk-quant"), the quantitative-programmer, the quantitative-trader or risk analyst—at a firm actively trading, risk managing, or even auditing books of derivatives—who needs to know the basic answer to this question, it is contained in a toolkit of well-established mathematical models and techniques. These professionals might be prepared to forgo mathematical rigor and the security against subtle errors that a thorough foundation could perhaps provide. They might be prepared to take a few shortcuts to get at these techniques relying on another team member or risk-group reviewer who has the thorough mathematical background to provide a safety net against subtle errors and misunderstandings. But the desk quant, the programmer, and the trader do not have to forgo everything. This book is aimed at such readers who want a good quick grasp of these techniques.

This first chapter reviews in coarse outline the two most typical techniques for theoretically modeling and pricing derivatives and takes the first motivational step of mentioning the model of the process for stock that underpins the first technique and the simplest implementations of the second. The remainder of the book will take the reader through the mathematical tools that underpin these two and many other techniques.

1.2. MODELS

1.2.1. What Is a Derivative?

The archetypal example of a *derivative* is a stock option. An equity call option is a financial contract. It is often an exchange traded security much like the stock itself. It is the right, but not the obligation, to buy one share of stock on a given date called the *expiration date* and at a price, the so-called *strike price,* contractually determined on the *trade date.* A put option is the right, but not the obligation, to sell at a preset *strike.* The action of buying or selling stock under the terms of a put or call option is called *exercise.* *American-style options* are exercisable on any day up to their expiration date—and *European options* are exercisable *only* on the expiration date.

Figure 1.1 shows the value of a call option on the expiration date, the *intrinsic value,* as a function of the market price of the stock (which we obviously don't know before expiration). Clearly, if the stock is trading above the strike, we can make a profit by exercising the option, that is, buying the stock for the strike price and then selling it in the marketplace for a profit of the difference between the two. If the stock price is below the strike on expiration date, the option is worthless. This graph of intrinsic value is obviously the fair price for the option at the moment of expiration. The fair price is generally different at all times previous to this. For example, it will have a small but nonzero fair value below strike at all times previous to expiration, representing the real, albeit small probability that the stock might end up above the strike at expiration.

Generally, a financial derivative or, better, a *contingent claim* is a loosely defined term meaning a security whose price is dependent on the price of another security that itself is called the derivative's *underlying* security. There may be more than one underlying. In the general case of an option, because there is some kind of choice involved, we can generally say that there clearly must be at least two underlying securities.

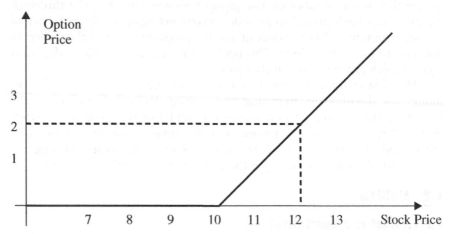

FIGURE 1.1   The value of a call option struck at 10 on maturity date as a function of the underlying stock price at maturity.

NOTE ON DERIVATIVE

The reader should beware of confusing the unrelated concept from mathematics, the derivative of a function, which means the rate of change of the given function with respect to one of its variables. Derivative in this mathematical sense can naturally arise in a discussion of financial derivatives, possibly even, and most confusingly, in the same sentence!

1.2.2. What Is a Model?

The typical mathematical pricing formula for options is the *Black-Scholes formula* (Figure 1.2). It outputs the theoretical fair price that a European option on a dividendless stock should trade for given the following set of inputs: call or put; current stock price; option strike price; time to expiration; average future stock volatility (for example, standard deviation of daily stock returns) from today to expiration, and current market value of the "riskless" interest rate from today to maturity (as determined by prices of treasury bonds or maybe interest rate swaps offered by big banks).

Deriving this formula requires a set of assumptions that are very important to bear in mind when using the formula. They are discussed in detail throughout this book. Essentially these assumptions are the *model*.

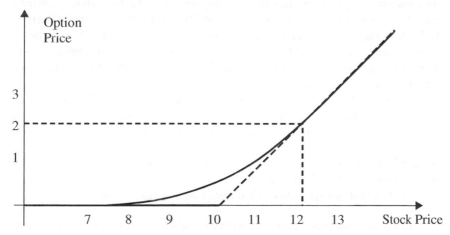

FIGURE 1.2   The value of a call option struck at 10 before maturity date as a function of the underlying stock price today.

Different sets of assumptions are different models. The model that underpins the Black-Scholes formula includes the following assumptions:

- Stock price changes are statistical (or stochastic), meaning the changes are random but they have a well-defined distribution of values.
- Stock price changes are continuous. The shorter the time over which you measure the average variation of size of changes, the smaller the average.
- The distribution of stock price changes is only a function of today's stock price and other values that are determined today and these changes have no dependence on the history of the stock or any other variables (models with this property are referred to as *Markovian* models)
- Stock price returns (i.e., changes expressed as percentages) have a normal (i.e., Gaussian) or bell-curve distribution around a mean that is the stock percentage growth rate.

1.2.3. Two Initial Methods for Modeling Derivatives

Many options and financial optionality can be approximately, quantitatively, and certainly qualitatively understood using simple modifications of the Black-Scholes formula. If this technique is lacking, the general machinery used to derive the formula provides a framework for more complex and thus more realistic analysis.

This means that an initial understanding of stock options and the Black-Scholes formula immediately opens up the world of derivatives and not just stock options themselves. The severe restriction is that the modeler can only study options under the assumption that the underlying security price has (continuous Markovian) normally distributed returns. Further study, leading to an understanding of the derivation of the Black-Scholes formula, opens that world far wider to practitioners, allowing them outside of the assumptions of a lognormal distribution of security changes to other perhaps more realistic distributions.

Mastering these two tools gives practitioners a strong suite of skills to analyze the trading and risks of derivatives. The derivatives accessible with the study of stock options include, among many others:

- Bond options and interest rate options such as swaptions
- Callable and putable bonds
- Credit default swaps (also called *credit-default put options*)

Indeed, any financial contract with an embedded clause that the issuer, or holder has the right but not the obligation to exercise can potentially be analyzed to some degree, even if only qualitatively, using one of the two methods mentioned: (1) the Black-Scholes formula with modifications or (2) the more general ideas behind the derivation of Black-Scholes formula.

1.2.4. Price Processes

The central idea behind the financial models that price stock options is the future distribution of stock prices. Furthermore, a random walk has characteristics similar to a stock or bond price path and has a future distribution of path end points. The walk may tend in a general direction, a *nonstochastic* drift, such as making a general observation that "on average" prices increase by $1 every year or each stock makes a $1/365 step up per day. But the walk also includes a *stochastic* component, referring to the fact that any particular price path has an additional random step either up or down.

There are many possible choices for this distribution of price changes and empirical data from the markets can often be inconclusive (principally because it seems the real world process itself is not stationary). More confusingly, the mathematical form of this distribution might look very different on different time scales. However, the normal distribution is a very useful tool for *first approximation* because it turns out that if the distribution of one-day price changes (i.e., steps) is normally distributed and has a standard deviation of, say, 5¢ with a mean of 0.27¢, then the distribution of annual changes (after many daily steps are added up) is also a normal distribution just with different mean and standard deviation. In fact, the normal distribution of annual price changes has a standard deviation of $0.05 \times \sqrt{365} \approx \$1$ and a mean of $\$0.0027 \times 365 \approx \$1$. The square root appeared because for normal distributions, the variance, the square of the standard deviation, grows linearly with time rather than the standard deviation itself.

This is an example of a model for the price process that starts at today's stock price; drifts by an average of $1 every year (and thus 0.27¢ every day); and has a standard deviation of annual changes of $1 (equivalent to a standard deviation of daily changes of 5¢).

A normal distribution of stock price changes with a drift could be used to find derivative prices with qualitatively correct features. The serious drawback, however, is that no matter where the stock starts, future stock prices have a potentially very large probability of being negative. This is a bad feature and will ruin even some qualitative observations, let alone quantitative ones. We need to refine this model slightly to get the simplest qualitatively correct model and hence make it (possibly) useful quantitatively even if only approximately.

1.2.5. The Archetypal Security Process: Normal Returns

This problem—of the normal distribution of future stock prices giving a probability for negative prices in the future—is easily corrected by modeling the price process with a normal distribution of returns on stock rather than the stock price changes. This is because allowing all possible percentage changes never results in negative stock prices if the path starts at a positive stock price. Equivalently, we may say that the natural log of stock prices is the normally distributed *state variable* over which the random walk occurs. This is a simple variable change and results in a much more compelling model qualitatively.

A normal distribution for changes of the natural log of stock price is equivalently described as a *lognormal* distribution of stock price changes (note this is a definition).

So one of the simplest models of the stock price process is given by a process in which the natural log of stock prices follows a random path, whose expected mean $\langle S \rangle$, grows with time (i.e., drifts) according to

$$\langle S \rangle = S_0 \exp(\mu(t - t_0)),$$

where the path begins at stock price S_0, at time t_0, and grows exponentially with time t according to some constant μ.

The path's size of statistical fluctuations is measured approximately by the standard deviation of the stock price percentage returns or, more accurately, by the standard deviation of the changes of the natural log of stock prices. This calculated result for an actual stock (or indeed any security) price history is common in derivatives finance and is called the *historical volatility* of the security.

$$\text{Historical Volatility} = \sqrt{\frac{\sum_{j=1}^{j=N}\left[\ln\left(\frac{S_j}{S_{j-1}}\right) - \text{mean}\left(\ln\left(\frac{S_j}{S_{j-1}}\right)\right)\right]^2}{N-1}}.$$

Note that $\ln(x)$ denotes the natural log of x, meaning that if $y = \ln(x)$ then $x = e^y$.

To bring the model we are constructing into contact with reality, imagine looking at a stock price that follows a path with a distribution of price changes that is *perfectly* lognormal.

First, for a large number of observations of price changes N, the average of the percentage changes of the stock price tends to a constant, the *expectation value,* driven by the model drift, reflecting that average prices in this model drift upward exponentially.

Second, the measured historical volatility will tend to the input volatility σ of the model as we calculate it using more and more stock price changes, that is, as N tends to infinity. Taking an average of many measurements n, of the historical variance over N stock price changes will have a distribution with a mean that is the input volatility squared. Alternatively, taking an average of n values for historical volatility using N market-close-price changes for the stock (*N-day historical vol'* in market parlance) will have a mean that is only approximately the input volatility, even as the number of observations n goes to infinity. However, as long as more than thirty price changes are used ($N > 30$) for each calculation of historical volatility, the difference between the expected value of the average and the input volatility is less than a few percent.

Note that an N-day historical volatility has an expected standard deviation of order volatility over $\sqrt{N}$. This means that an observed path of trailing historical volatility fluctuates around its mean with percentage difference to the mean of order $1/\sqrt{N}$. For example, a graph of trailing 30-day historical volatility on a perfectly lognormal stock price with actual volatility 0.30, will fluctuate around a mean value within a few percent of 0.30, and more than half the time will be within a band approximately given by 0.25 and 0.35. In conclusion, we *can* make observations directly comparing or contrasting real markets with this lognormal model.

These features of a lognormal distribution are already of use for some qualitative match to the real markets. For example, historical graphs of the major indices show long periods of something like exponential growth (most easily seen as straight lines on *log index* versus *time* plots) and graphs of trailing historical volatility, for a particular stock, do show a lower variance with larger samples of daily price changes. However, while a cursory look at the data shows some qualitative match, it simultaneously shows significant differences that make this model a schematic fit at best. Typical stocks can have a historical volatility graph that has a mean near 0.30 and typical range of fluctuations between 0.25 and 0.35 for a year or more and then a business announcement can cause a stock price change that makes historical volatility jump to 0.60. Such events are too frequent for the lognormal distribution and lead directly to discussions of fat tails.

Finally, if this process were to imply that there is a unique option price, then this would obviously be a very valuable qualitative tool with the potential to be quantitatively useful. This theoretical option price might not match the market exactly because the real market for the stock does not have exactly lognormally distributed stock price changes (in fact, the real market is not continuous and is influenced by the recent past, i.e. non-Markovian), but this price would still be a useful guide to very expensive or very cheap options and how to trade and risk manage them.

Many processes for stocks can indeed be used to derive option prices and the simplest model—an option on a stock with a lognormally distributed price process and no dividend—results in the famous Black-Scholes formula for the price of the option. Understanding the techniques of modeling price processes and of finding option prices consistent with a chosen underlying price process are the principal objectives of this book.

1.2.6. Book Outline

This book outlines all of the basic mathematics to understand the derivation of pricing equations for various derivatives from the starting point of assuming a process for the underlying security price. It turns out that a partial differential equation (PDE) exists for each process and which the pricing formula of all derivatives on this underlying security solve. We will derive this equation and solve it. Solving the equation can be done analytically or, more likely, numerically.

After some preliminary mathematics review in chapter 2, chapters 3 and 4 deal with a formalism to describe stochastic processes and its application to finance. Chapters 5 and 6 then derive the pricing equation based solely on stochastic price fluctuations as the source of risk. Chapters 7 and 8 repeat the development again for interest rates with a focus on the constraints on the form for the process that arise due to trying to consistently model the stochastic movements of the interest rate *curve* rather than just a single security price. Chapter 9 deals with analytic and numerical solutions to the types of equations that arise in derivatives modeling. Chapter 10 incorporates a simple binary probabilistic model of default into the framework and finally Chapter 11 combines much of the previous work into some specific examples of models of derivative securities. A few sample exercises together with solutions are provided and range from the almost inane, but actually very important, *"do I get this?"* type question to much more difficult *"I can't do this!"* type questions. Four topics are relegated to appendices because they were either algebra intensive or outside the flow of the text but nevertheless they are important.

CHAPTER 2

Preliminary Mathematical Tools

2.1. PROBABILITY DISTRIBUTIONS

A review of discrete and continuous probability distributions is an appropriate first step to outline the mathematics of derivatives.

If p_i (or $P(x)dx$) represents the probability of an "occurrence" at discrete coordinate x_i (or in range $x \to x + dx$), then we can model a person's ability to throw darts, say, by the probability distribution of the results of their throwing darts at a board, aiming to hit only the bull's-eye. The distribution would likely be angularly symmetric and would peak around the bull's-eye and fall off farther away. This means that the thrower has a good enough eye that his throws usually come close to the bull's-eye. We might even measure his skill level by trying to measure the distribution.

Presumably, the average result of many throws

$$\langle x \rangle = \sum_i x_i p_i \quad or \quad \langle x \rangle = \int_{x-space} x P(x)\,dx$$

will be close to the bull's-eye, and would get closer and closer as more and more throws are observed and so contains little or no information about the thrower's skill level. On the other hand, the width of the distribution (or radius of the central lump of distribution of throws) can be measured as the standard deviation σ where

$$\sigma^2 = \langle (x - \langle x \rangle)^2 \rangle = \langle x^2 \rangle - \langle x \rangle^2 = \sum_i x_i^2 p_i - \left(\sum_i x_i p_i \right)^2$$

$$or\ \sigma^2 = \langle x \rangle = \int_{x-space} x^2 P(x)\,dx - \left(\int_{x-space} x P(x)\,dx \right)^2.$$

11

EXPECTATION VALUES AND AVERAGES

The average of N values $x_1, x_2, \ldots, x_N$, is often written as

$$\bar{x} = \frac{1}{N} \sum_{j=1}^{j=N} x_j.$$

Now if the N values are selected from a randomly distributed variable x, with continuous probablity distribution $p(x)dx$, then the expectation value of x is written

$$\langle x \rangle = \int_{-\infty}^{\infty} x p(x) dx.$$

The value $\bar{x}$ is itself stochastic and has an expectation value and variance itself. The important idea is that this expectation value is the limit of $\bar{x}$ as N tends to infinity. Furthermore, the variance of x and goes to zero as N tends to infinity. To avoid this subtle notation, the mean (and variance) are often written out in full in this text,

$$\text{mean}(x) = \frac{1}{N} \sum_{j=1}^{j=N} x_j,$$

while the expectation value $\langle x \rangle$, is used to denote the expectation value of the mean.

This is a smaller and smaller number depending on the thrower's level of skill. The more accurate the thrower, the tighter the distribution around the bull's-eye, and the higher the probability of hitting the bull's-eye with any one throw. This is possibly a good measure of skill level. We might imagine that it is approximately constant but slowly changes as a particular thrower's skill level improves (or weakens) with practice (or lack of practice).

For distributions then, we may ascertain the expectation value of the coordinate x, and indeed functions of the coordinate $f(x)$. Also note that a "state-density" function, w_i, or $w(x)$, may be included.

$$\{x_i, p_i\}: \quad \langle x \rangle = \sum_i x_i p_i, \quad if \ \sum_i p_i = 1 \ and \ p_i > 0, \ \forall(i)$$

$$\{x_i, p_i, f(x_i)\}: \quad \langle f(x) \rangle = \sum_i f(x_i) p_i, \quad if \ \sum_i p_i = 1$$

$$\{x_i, p_i, w_i, f(x_i)\}: \quad \langle x \rangle = \frac{\sum_i x_i w_i p_i}{\sum_i w_i p_i}, \quad \langle f(x) \rangle = \frac{\sum_i f(x_i) w_i p_i}{\sum_i w_i p_i}$$

$$\{x, P(x), w(x)\}: \quad \langle x \rangle = \int_{x-space} f(x) P(x) w(x)\, dx, \quad if \int_{x-space} P(x) w(x)\, dx$$

$$= 1$$

A Gaussian distribution (also called a normal distribution or bell-curve distribution) is a very important example of a distribution. Another important example is the lognormal distribution. One is a "variable change" of the other. To execute variable changes for distributions (say $x \to y$), we can replace the variable function in the function $p(x)$ as follows:

$$p'(y) = p(x(y)),$$

but it is equally important to transform the volume element dx (under the integral):

$$dx = J(y)\, dy.$$

This function J is called a *Jacobian* and has various forms for various dimensions, that is, variable changes from 1 to 1 dimension, or 2 to 2 dimensions, or 3 to 3 dimensions and so on. This function can be understood as a "weight" function, a necessary term to understand variable changes under the integral. For example, the Jacobian for 1 dimension (1-D) is

$$dx = \frac{\partial x}{\partial y}\, dy.$$

For higher dimensions we will need a special algebra to calculate the Jacobian easily; the algebra of 1-forms and n-forms.

First, let's work on a simple (1-D) example and calculate the normalization integral for a typical Gaussian distribution.

$$p(x) = \exp\left(-\frac{x^2}{2}\right)$$

$$I = \int_{-\infty}^{\infty} p(x)\, dx$$

We are going to consider the value of I^2 and then make a 2-D variable change

$$x = r\cos\vartheta$$

$$y = r\sin\vartheta$$

that results in

$$I^2 = \int \int dx\,dy\,p(x)p(y) = \int \int dr\,d\vartheta \frac{J(x,y)}{J(r,\vartheta)}p(x(r,\vartheta))p(y(r,\vartheta))$$

$$= \int \int dr\,d\vartheta \frac{J(x,y)}{J(r,\vartheta)}p(r).$$

This requires the calculation of the 2-D Jacobian, which is given by

$$\frac{J(x,y)}{J(r,\vartheta)} = \frac{\partial x}{\partial r}\frac{\partial y}{\partial \vartheta} - \frac{\partial y}{\partial r}\frac{\partial x}{\partial \vartheta} = r.$$

We can now integrate over the angle ϑ (the argument does not depend on angle) to get a factor of 2π and then note, using the chain rule, that

$$\frac{\partial p(r)}{\partial r} = -rp(r),$$

for $p(x)$ defined above, to find

$$I^2 = \int \int dr\,d\vartheta\,rp(r) = 2\pi \int dr\,rp(r) = 2\pi[-p(r)]_{r=0}^{r=\infty}$$

$$= 2\pi \left[-\exp\left(-\frac{r^2}{2}\right) \right]_{r=0}^{r=\infty} = 2\pi.$$

So the result is $I^2 = 2\pi$ and the correctly normalized distribution is given by

$$p(x) = \frac{1}{\sqrt{2\pi}} \exp\left(-\frac{x^2}{2}\right).$$

It remains only to note that the mean and standard deviation of this Gaussian distribution have particular values, as shown in the exercises that follow the next section.

2.2. n-DIMENSIONAL JACOBIANS AND n-FORM ALGEBRA

The 2-D Jacobian is given by

$$\frac{J(x_1, x_2)}{J(y_1, y_2)} = \frac{\partial x_1}{\partial y_1}\frac{\partial x_2}{\partial y_2} - \frac{\partial x_2}{\partial y_1}\frac{\partial x_1}{\partial y_2}.$$

Indeed the n-dimensional Jacobian is given by

$$\frac{J(x_1, x_2, \ldots, x_N)}{J(y_1, y_2, \ldots, y_N)} = \det \begin{bmatrix} \frac{\partial x_1}{\partial y_1} & \frac{\partial x_1}{\partial y_2} & \cdots & \frac{\partial x_1}{\partial y_N} \\ \frac{\partial x_2}{\partial y_1} & \frac{\partial x_2}{\partial y_2} & \cdots & \frac{\partial x_2}{\partial y_N} \\ \vdots & \vdots & & \vdots \\ \frac{\partial x_N}{\partial y_1} & \frac{\partial x_N}{\partial y_2} & \cdots & \frac{\partial x_N}{\partial y_N} \end{bmatrix}.$$

The generally much easier method to get the result, for a specific case instead of general functions, is to use n-form algebra. The rules of this algebra follow.

If x and y are scalars (i.e., ordinary numbers that are only magnitude and are not associated with a direction) then dx, dy, and so on (i.e., the line elements) are "1-forms." They are *not* scalars (indeed, they are associated with a direction) and we need to know their algebra rules. In fact, they anticommute but are distributive (the following expressions include first a 2-form and second, another 1-form):

$$dx \, dy = -dy \, dx$$

$$dx + dy = dy + dx$$

Sometimes this is written

$$dx \wedge dy = -dy \wedge dx$$

$$dx + dy = dy + dx$$

using a special symbol for *form* multiplication, because an n-form times an m-form results in an $(n+m)$-form. 1-forms commute with regular numbers (e.g., the scalar function $f(x, y)$, a 0-form) but cannot be added to them.

$$f \, dy = dyf$$

$$f + dx \cdots meaningless.$$

The last expression is "adding a vector to a scalar", and this has no meaning. Indeed, addition is meaningful only among forms of the same order. Finally, the scalar functions $f(x, y)$ and $g(x, y)$ can generate 1-forms as follows.

Scalars are expressed as

$$f(x, y), \ g(x, y);$$

1-forms as

$$df = \frac{\partial f}{\partial x} \, dx + \frac{\partial f}{\partial y} \, dy,$$

$$dg = \frac{\partial g}{\partial x} \, dx + \frac{\partial g}{\partial y} \, dy;$$

and 2-forms as

$$d^2f = d(df) = \frac{\partial^2 f}{\partial x \partial y} dydx + \frac{\partial^2 f}{\partial x \partial y} dxdy = \left(\frac{\partial^2 f}{\partial x \partial y} - \frac{\partial^2 f}{\partial x \partial y} \right) dydx = 0,$$

$$d^2g = 0,$$

$$df\,dg = \left(\frac{\partial f}{\partial x} \frac{\partial g}{\partial y} - \frac{\partial f}{\partial y} \frac{\partial g}{\partial x} \right) dxdy.$$

This last line gives the 2-D Jacobian; this is the machinery we need to perform variable changes on volume elements. The value of this n-form algebra is that it becomes very easy for specific examples:

$$x = r\cos\vartheta, \quad y = r\sin\vartheta;$$

$$r = \sqrt{x^2 + y^2}, \quad \vartheta = \tan^{-1}\left(\frac{y}{x}\right);$$

$$dx = \cos\vartheta\,dr - r\sin\vartheta\,d\vartheta,$$

$$dy = \sin\vartheta\,dr + r\cos\vartheta\,d\vartheta,$$

$$dxdy = (r\cos^2\vartheta + r\sin^2\vartheta)\,dr\,d\vartheta = rdr\,d\vartheta;$$

$$\text{i.e.,} \quad \frac{J(x,y)}{J(r,\vartheta)} = r$$

Thus we can directly relate the 2-D volume elements under a change of variables and then read off the Jacobian easily.

2.3. FUNCTIONAL ANALYSIS AND FOURIER TRANSFORMS

Functional analysis or distribution theory deals with integrals on continuous distributions. Here we note that a wider definition of integration is needed (rather than the usual Riemann) and this is Lebesgue integration. Riemann integration is the discrete sum of small increments times function values with the increment size tending to zero and therefore the number of increments tending to infinity. If this limit is well-defined, and the result is independent of the method of shrinking the increments, then this limiting result is the Riemann integral value.

Lebesgue integration works over a wider class of integrands (called *distributions*) than Riemann integration that only works on smooth functions. This is because a parameter is introduced and the Riemann integral is performed on the function for a particular value of the parameter, which then is allowed to go to a limit *after* integration. This limit might be a value of the parameter for which the Riemann integral is not defined while the limiting value itself is defined. This limiting value is then the value of the Lebesgue integral.

An example is much easier to digest—consider the Dirac delta. If we look at the full normal distribution formula, parameterized by the mean and standard deviation, μ and σ,

$$p(x)\,dx = \frac{1}{\sqrt{2\pi\sigma^2}}\exp\left(-\frac{(x-\mu)^2}{2\sigma^2}\right)dx,$$

we may note that this function is not defined in the limit $\sigma \to 0$. But generally Riemann integrals of the form

$$I[f] = \int_{-\infty}^{\infty} f(x)p(x)\,dx$$

are well defined for all $\sigma > 0$ (but not 0). For example, if $f(x) = 1, I[f] = 1$. Furthermore, for $f(x) = x, I[f] = \mu$ and also for $f(x) = (x - \mu)^2, I[f] = \sigma^2$ and so on.

The Dirac delta, $\delta(x - \mu)$, is defined as the "distribution" represented by the limit $\sigma \to 0$ for the function $p(x)$, with this limit taken after integration.

It immediately follows that

$$\int_{-\infty}^{\infty} \delta(x - \mu)\,dx = 1,$$

$$\int_{-\infty}^{\infty} \delta(x - \mu)f(x)\,dx = f(\mu)$$

where now the integral is understood to be a Lebesgue integral. We see that the Dirac delta is a function that is zero everywhere but infinitely high at the origin (loosely speaking) and has an area underneath it of 1. Mathematically we can define such a function; but Riemann integration of it is certainly not defined because it is not smooth enough but a "distribution" *is* well defined (i.e., very loosely "a function under the integral").

Now consider the Fourier transform of $f(x), f(k)$ say,

$$f(k) = \frac{1}{\sqrt{2\pi}}\int_{-\infty}^{\infty} f(x)e^{-ikx}\,dx$$

$$f(x) = \frac{1}{\sqrt{2\pi}}\int_{-\infty}^{\infty} f(k)e^{ikx}\,dk$$

where Lebesgue integration is a requirement to ensure that "Fourier analysis" makes sense, and the second relation representing the inverse is included. (Here $i = \sqrt{-1}$.)

Then we find for a normal distribution with mean μ and variance σ^2,

$$p(x) = \frac{1}{\sqrt{2\pi\sigma^2}} \exp\left(-\frac{(x-\mu)^2}{2\sigma^2}\right)$$

$$p(k) = \frac{1}{\sqrt{2\pi}} \exp\left(\frac{-k^2\sigma^2}{2}\right) e^{-ik\mu},$$

the Fourier transform is *also* a normal distribution. But note that if the peak of the original normal distribution is thin, that is, $\sigma \to 0$, then the spectrum distribution, $p(k)$, is wide, and vice versa.

Now consider *convolution*: it looks like multiplication in Fourier space. Define $h = f * g$ (here the * denotes the operation of convolution):

$$h(x) = [f^*g](x) = \int_{-\infty}^{\infty} f(y-x)g(y)\,dy.$$

Then take the Fourier transform of $h(x)$ and it turns out:

$$h(k) = f(k)g(k).$$

It is a very simple result. The spectral distributions multiply to convolute the functions themselves. This is a powerful tool.

Finally note that for functions $f(x)$ and $g(x)$ together with their Fourier transforms, $f(k)$ and $g(k)$, then if

$$g(x) = \frac{df(x)}{dx},$$

then

$$g(k) = ikf(k),$$

and again in Fourier transform space, the operation of taking the first derivative is simplified to multiplication by the spectral variable and i.

2.4. NORMAL (CENTRAL) LIMIT THEOREM

Any "shock" (over a constant state density), repeated, eventually becomes a normal distribution if the variance of the shock is finite. This may be mathematically described as an initial distribution $g(x)$, which is convoluted with a shock distribution $f(x)$ to produce a new distribution $h(x)$:

$$h(x) = [f * g](x) = \int_{-\infty}^{\infty} f(y-x)g(y)\,dy$$

and this result is again convoluted with the shock and so on many times. Indeed it may also be thought of as the convolution of the shock distribution $f(x)$ with *itself* many times, and this result is convoluted with the initial distribution

$$f^N(x) = [f * f * f \cdots * f](x)$$

giving a final result

$$Final(x) = \left[f^{(*N)} * g\right](x).$$

It turns out that under the conditions that the shock is normalized to 1, and the mean and variance of the shock $f(x)$ are well-defined (say mean μ and variance σ^2) that

$$f^N(x) \underset{N \to \infty}{\to} \frac{1}{\sqrt{2\pi\sigma^2 N}} \exp\left(-\frac{(x - \mu N)^2}{2\sigma^2 N}\right).$$

Thus any shock or smearing out of a function, if repeated over and over, looks like a normal distribution convoluted with the initial distribution. This is the central limit theorem: for a plausibility argument see Appendix A.

A specific example would be a 1-D random walk in which the walker starts at the origin; an initial distribution corresponding to a Dirac delta. The walker then takes a step right 1 unit or a step left 1 unit, each with probability 50%; this is a shock distribution that is a sum of two Dirac deltas multiplied by $\frac{1}{2}$, one at $x = -1$, the other at $x = +1$, as in Figure 2.1.

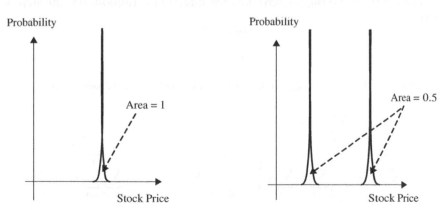

FIGURE 2.1 Schematic plots of Dirac delta functions representing the initial distribution and shock distribution of a path. (a) Initial stock distribution; (b) shock distribution

This shock distribution has a mean μ of zero and variance σ^2 of 1 and, using the central limit theorem, has a large number of steps N limit (many convolutions of the shocks in Figure 2.1) given by

$$p(x) \approx \frac{1}{\sqrt{2\pi N}} \exp\left(-\frac{(x)^2}{2N}\right).$$

The next section examines random walks in more detail.

2.5. RANDOM WALKS

Define a random walk in 1-D, or 2-D and so on, as the result X_N, of N steps:

$$X_N = \sum_{i=1}^{N} x_i$$

where each x_i is selected from a (possibly different) random distribution. The sums of random variables follow some rules. The "sum of Gaussians is a Gaussian" meaning if each x_i is selected from a normal—that is, a Gaussian—distribution, then the total, X_N, is also distributed according to a Gaussian. This is because convoluting Gaussians produces a Gaussian. Another rule is that the sum of variables obeying any other distribution on a constant state density goes to a Gaussian as the number of steps goes to infinity if the standard deviation is well defined (normal limit theorem).

Furthermore, carefully note that the standard deviation of a walk is different from the standard deviation of the mean. Let's say the average step is size 1 but the average is zero (i.e., the direction is random but the step is size 1),

$$\langle x_i \rangle = 0,$$

$$\langle x_i^2 \rangle = 1.$$

Expected (i.e., average) length of a walk and standard deviation is

$$\langle X_N \rangle = \sum_{i=1}^{N} \langle x_i \rangle = 0,$$

$$\sqrt{\langle (X_N)^2 \rangle} = \sqrt{\left\langle \sum_{i=1}^{N} x_i \sum_{j=1}^{N} x_j \right\rangle} = \sqrt{\sum_{i=1}^{N} \sum_{j=1}^{N} \langle x_i x_j \rangle} = \sqrt{\sum_{i=1}^{N} \sum_{j=1}^{N} \delta_{ij}} = \sqrt{N}.$$

The Kronecker delta δ_{ij} has the value 1 if the indexes are the same and zero otherwise. It shows up due to the fact that the steps are uncorrelated, and so only the same step gives a contribution due to its variance of 1.

So the walk has step size 1, but after N steps the path has typically moved $\sqrt{N}$ steps from the origin, even though the average is near zero; that is, there is a wide distribution of results between $-\sqrt{N}$ and $\sqrt{N}$.

Now consider the average of N individual steps and its expectation value and standard deviation:

$$\left\langle \frac{X_N}{N} \right\rangle = \frac{\sum_{i=1}^{N} \langle x_i \rangle}{N} = 0,$$

$$\sqrt{\left\langle \left(\frac{X_N}{N}\right)^2 \right\rangle} = \sqrt{\frac{\sum_{i=1}^{N} x_i}{N} \frac{\sum_{j=1}^{N} x_j}{N}} = \frac{\sqrt{\sum_{i=1}^{N} \sum_{j=1}^{N} \delta_{ij}}}{N} = \frac{1}{\sqrt{N}}.$$

Thus the "error" in the average (its standard deviation) tends to zero as $N \to \infty$.

Now, if x is selected from a normal distribution with mean μ and variance σ, and so

$$p(x)\,dx = \frac{1}{\sqrt{2\pi\sigma^2}} \exp\left(-\frac{(x-\mu)^2}{2\sigma^2}\right) dx,$$

then the distribution of $X_N = \sum_{i=1}^{N} x_i$, is

$$p(X_N)dX_N = \frac{1}{\sqrt{2\pi N\sigma^2}} \exp\left(-\frac{(X_N - N\mu)^2}{2N\sigma^2}\right) dX.$$

Also note

$$\langle \exp(x) \rangle = \exp\left(\langle x \rangle + \frac{1}{2}\mathrm{var}(x)\right) = \exp\left(\mu + \frac{\sigma^2}{2}\right)$$

and if $X_N = \sum_{i=1}^{N} x_i$ so that X_N is a sum of Gaussians, then

$$\langle \exp(X_N) \rangle = \exp\left(N\langle x \rangle + \frac{1}{2}N\mathrm{var}(x)\right) = \exp\left(N\mu + \frac{N\sigma^2}{2}\right).$$

We have already obtained a result that has a financial interpretation: Reinvested capital, due to compounding, results in monetary risk (i.e., variance) being rewarded on average with excess returns. If the returns on investments are normally distributed, and if average returns are $\langle X_N \rangle = rT$ while the variance (or risk) of these returns is $\langle X_N^2 \rangle - \langle X_N \rangle^2 = \sigma^2 T$ then the average compounded returns are

$$\langle \exp(X_N) \rangle = \exp\left(rT + \frac{\sigma^2 T}{2}\right).$$

2.6. CORRELATION

For numbers selected from random distributions in dimensions greater than one, we can consider correlations of the variables selected at the same time to be nonzero. Correlation is defined as

$$\rho = \frac{\langle xy \rangle - \langle x \rangle \langle y \rangle}{\sigma_x \sigma_y}.$$

This is the same definition for continuous or discrete distributions. The value of correlation is always between -1 and 1. The formula makes more intuitive sense rearranged as

$$\langle xy \rangle = \langle x \rangle \langle y \rangle + \rho \sigma_x \sigma_y,$$

which means the probability of two events x and y, occurring simultaneously, is just the product of their individual (unconditional) probabilities of occurrence only if they are uncorrelated, that is, $\rho = 0$, and the probability of coincidence is higher than the unconditional coincidence if the events are correlated, that is, $\rho > 0$, and it is lower if they are anti-correlated, namely, $\rho < 0$.

We may construct a correlated system out of an uncorrelated one: if z_1 and z_2 are uncorrelated variables selected from Gaussian distributions with mean zero and variance 1,

$$p(z_i)dz_i = \frac{1}{\sqrt{2\pi}} \exp\left(-\frac{z_i^2}{2}\right) dx,$$

and then the system of two variables has a distribution function,

$$p(z_1, z_2)dz_1 dz_2 = \frac{1}{\sqrt{2\pi}} \exp\left(-\frac{z_1^2}{2}\right) \frac{1}{\sqrt{2\pi}} \exp\left(-\frac{z_2^2}{2}\right) dz_1 dz_2.$$

Also note that

$$\rho_{12} = \langle z_1 z_2 \rangle = \int_{-\infty}^{\infty} dz_1 \int_{-\infty}^{\infty} dz_2 z_1 z_2 p(z_1, z_2) = 0.$$

Now change variables,

$$x_1 = \alpha_1(z_1 + \beta z_2) + \mu_1$$
$$x_2 = \alpha_2(z_1 - \beta z_2) + \mu_2.$$

We are free to choose α_1, α_2 and β so that

$$\langle x_1 \rangle = \mu_1,$$

$$\langle x_2 \rangle = \mu_2,$$

$$\text{var}(x_1) = \sigma_1^2,$$

$$\text{var}(x_2) = \sigma_2^2,$$

$$\rho = \frac{\langle x_1 x_2 \rangle - \mu_1 \mu_2}{\sigma_1 \sigma_2},$$

given that

$$\langle z_1 \rangle = 0,$$

$$\langle z_2 \rangle = 0,$$

$$\text{var}(z_1) = 1,$$

$$\text{var}(z_2) = 1,$$

$$\langle z_1 z_2 \rangle = 0.$$

If the distribution of (z_1, z_2) is

$$p(z_1, z_2) dz_1 dz_2 = \frac{1}{2\pi} \exp\left(-\frac{z_1^2 + z_2^2}{2}\right) dz_1 dz_2,$$

then the distribution of (x_1, x_2) is

$$p(x_1, x_2) dx_1 dx_2$$

$$= \exp\left[-\left(\frac{\sigma_2^2(x_1 - \mu_1)^2 + \sigma_1^2(x_2 - \mu_2)^2 - 2\rho\sigma_1\sigma_2(x_1 - \mu_1)(x_2 - \mu_2)}{2(1 - \rho^2)\sigma_1^2\sigma_2^2}\right)\right]$$

$$\frac{dx_1 dx_2}{2\pi\sigma_1\sigma_2\sqrt{1 - \rho^2}}.$$

The conclusion here is always switch to (z_1, z_2) variables because it is much easier to work in (z_1, z_2) space than (x_1, x_2) space. Simply stated, when dealing with variables that have means, variances and correlations, switch to variables with means of 0, variances of 1, and correlations of 0.

Randomly distributed variables with a correlation can be interpreted as the linear sum of random variables with no correlation. Thus correlation measures the degree to which "mixing" occurs between the two variables, and the likelihood that the variables will move in a synchronized fashion; and the range of correlation from -1 to 1 naturally expresses this.

2.7. FUNCTIONS OF TWO/MORE VARIABLES: PATH INTEGRALS

Let's review. Given a well-defined function of two variables,

$$f(x, y),$$

we can write (using 1-form algebra)

$$df = \frac{\partial f}{\partial x} \, dx + \frac{\partial f}{\partial y} \, dy.$$

Across the 2-D x,y surface this can have an intuitive meaning as follows: If we chose a path, such that

$$x = x(t),$$

$$y = y(t)$$

for $0 \leq t \leq 1$, say, then we find that along this path

$$\frac{df}{dt} = \frac{\partial f}{\partial x} \frac{dx}{dt} + \frac{\partial f}{\partial y} \frac{dy}{dt}.$$

For example, define $f(x, y) = x^2 y^2$ and the path as follows:

$$x = \cos(\pi t),$$

$$y = \sin(\pi t)$$

for $0 \leq t \leq 1$ as plotted in Figure 2.2.

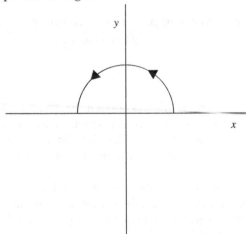

FIGURE 2.2 Plot of a path on the (x, y) plane given by $x = \cos(\pi t), y = \sin(\pi t)$ for $0 \leq t \leq 1$

Then

$$\frac{df}{dt} = -2xy^2\pi \sin(\pi t) + 2x^2 y\pi \cos(\pi t).$$

Specifically, along this path the value of $\frac{df}{dt}$ is defined. For another path this is incorrect. But the expression

$$df = \frac{\partial f}{\partial x} dx + \frac{\partial f}{\partial y} dy$$

encompasses all possible paths.

Another intuitive meaning for 1-forms is that they are "really" defined under the integral, and we see that along a given path

$$\int_{t=0}^{t=1} df = \int_{t=0}^{t=1} \frac{\partial f}{\partial x} dx + \frac{\partial f}{\partial y} dy$$

$$= \int_{t=0}^{t=1} \left(\frac{\partial f}{\partial x} \frac{dx}{dt} + \frac{\partial f}{\partial y} \frac{dy}{dt} \right) dt$$

$$= f(x(t=1), y(t=1)) - f(x(t=0), y(t=0)).$$

We can see that along a closed path—such as the path parameterization just displayed, except now $0 \leq t \leq 2$ and this closes the full circle of the path as in Figure 2.3—the integral is

$$\int_{t=0}^{t=2} df = \oint df = 0.$$

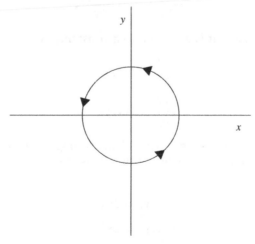

FIGURE 2.3   Plot of a path on the (x, y) plane given by $x = \cos(\pi t), y = \sin(\pi t)$ for $0 \leq t \leq 2$.

There is an important exception to this. Consider, after a change of variables along the same path, but importantly with a *different* function f:

$$x = r\cos\vartheta, \quad y = r\sin\vartheta;$$

$$r = \sqrt{x^2 + y^2}, \quad \vartheta = \tan^{-1}\left(\frac{y}{x}\right);$$

$$f(r, \vartheta) = \vartheta$$

$$r = 1, \quad \vartheta = \pi t; 0 \le t \le 2.$$

Here we find

$$\int_{t=0}^{t=2} df = \int_{\vartheta=0}^{\vartheta=2\pi} d\vartheta = 2\pi.$$

All that happened in this special case is that the closed path enclosed a single point where the function was not defined, at the origin. For all paths not enclosing this point the answer is still zero.

2.8. DIFFERENTIAL FORMS

Now we consider a more general 1-form:

$$df = h(x, y)\,dx + g(x, y)\,dy$$

Note that under the condition

$$\frac{\partial h(x, y)}{\partial y} = \frac{\partial g(x, y)}{\partial x},$$

then $f(x, y)$ exists, and it is given, up to a constant, by

$$\frac{\partial f}{\partial x} = h(x, y),$$

$$\frac{\partial f}{\partial y} = g(x, y)$$

and the path integral is path independent as long as the equivalent paths do not enclose singular points of the functions $h(x, y)$ and $g(x, y)$, using the previous example:

$$h(x, y) = 2xy^2,$$

$$g(x, y) = 2x^2 y,$$

$$\Rightarrow f = x^2 y^2.$$

CHAPTER 3

Stochastic Calculus

3.1. WIENER PROCESS

We have already looked at the concept of a random walk, starting at zero,

$$X = \sum_{i=1}^{N} \varepsilon_i.$$

The result of N steps of size 1 in a randomly selected direction gets the walker to position X; the mean of X is zero and the standard deviation of X is $\sqrt{N}$ in the large N limit. The same is true for any shock of the same standard deviation and mean, for example, if each ε_i is selected from a normal distribution of standard deviation 1 and mean 0.

In this chapter, we modify the analysis slightly. We will make the steps occur every increment of time Δt, and the step is going to be selected from a random distribution of mean zero and standard deviation $\sqrt{\Delta t}$,

$$z_i = \sqrt{\Delta t}\varepsilon_i,$$

$$\Delta t = \frac{T}{N}.$$

Then define

$$X_T = \lim_{N \to \infty} \sum_{i=1}^{N} z_i.$$

This formula is mathematically well defined for any finite N; because the behavior is smooth over large values of N, the limit for $N \to \infty$, is well-defined as well. Visually we have a wiggly path. But it is impossible to correctly draw! This is because it has a fractal nature to it. No matter how small you go, it still looks similar, and the derivative is *not* defined no matter how small the line element you take.

The above (limiting) expression will be codified by the following expressions:

$$X_T = \int_{t=0}^{t=T} dz(t),$$

$$dX = dz(t)$$

These expressions are shorthand for the well-defined version displayed previously. We have an expectation value under the probability distribution of all increments:

$$\langle dz(t) \rangle = 0,$$

$$\langle dz(t)^2 \rangle = dt,$$

$$\langle dz(t)\, dz(s) \rangle = dt\, ds\, \delta(t - s)$$

We may state this as the stochastic increments have mean zero, standard deviation dt, and at different times are all uncorrelated. These three expressions are the basic tools to manipulate stochastic integrals and then take expectation values of the results. For instance, these expressions imply (more strictly: *are implied by!*)

$$\langle X \rangle = 0,$$

$$\langle X^2 \rangle = T.$$

Last, we can construct a new variable X of arbitrary volatility σ (i.e., standard deviation per square root of time period) and drift μ (i.e., rate of change of mean per time period):

$$X_T = \int_{t=0}^{t=T} dX,$$

$$dX = \mu(t)\, dt + \sigma(t)\, dz(t)$$

and then

$$\langle X_T \rangle = \int_{t=0}^{t=T} \mu(t)\, dt,$$

$$\langle X_T^2 \rangle - \langle X_T \rangle^2 = \int_{t=0}^{t=T} \sigma^2(t)\, dt.$$

This is called a Brownian motion. It can be generalized to the following form:

$$dX = \mu(X, t)\, dt + \sigma(X, t)\, dz(t).$$

Note the "stochastic-ness" of the process is contained solely in the $dz(t)$ term. At time t, the values of the drift rate and volatility are well defined; the dt term integrates up to a regular function while the $dz(t)$ term contains the stochastic character of the process for X. The simplest Brownian motion is of annual variance 1, and mean 0,

$$dX = dz(t)$$

and is called a *Wiener process*.

The distribution and specifically means and variances of the path endpoints for this more general process are harder to calculate; but the definition of standard deviation for a path looks at the changes in the path from sample time to sample time. Given an $N + 1$ time series of path values for X, $(r_1, r_2, \ldots, r_{N+1})$ say, this generates N samples of changes, and a sample of variances:

$$\text{Mean} = \langle \overline{\Delta r} \rangle = \frac{1}{N} \sum_{i=1}^{N} \Delta r_i = \frac{1}{N} \sum_{i=1}^{N} (r_{i+1} - r_i),$$

$$\text{Variance} = \langle (\Delta r - \langle \overline{\Delta r} \rangle)^2 \rangle = \frac{1}{(N-1)} \sum_{i=1}^{N} (\Delta r_i - \langle \overline{\Delta r} \rangle)^2,$$

$$\text{Standard deviation} = \sqrt{\text{variance}}.$$

The $N - 1$ factor shows up because the expression then gives a better estimate of the variance for small N (this can be derived). Note that it gives infinity for $N = 1$, meaning, if we have only two stock price points, then we have a single estimate of the mean return, but no error on the mean return, because we have only one sample of this. The standard deviation is the square root of the variance.

Clearly as $N \to \infty$, and the time-step goes to zero (and for a particular path value, X, and particular time, t), we would expect to get:

$$\text{Standard deviation} = \sigma.$$

Note that in reality it is going to be difficult to measure if there is time dependence of the standard deviation term—and very difficult if there is path value dependence. It would be necessary to perform multiple measurements of path value changes at nearly the same time and at very similar values for the path.

3.2. ITO'S LEMMA

The lemma can be proven for a stochastic process using the definition of a Wiener process,

$$x_T = \lim_{N \to \infty} \sum_{i=1}^{N} z_i,$$

$$z_i = \sqrt{\Delta t}\varepsilon_i$$

where ε_i is selected from a normal distribution of variance 1 and mean 0, which is written in shorthand as

$$dx = dz(t).$$

Then consider the process for a function of this stochastic variable $f(x)$. The result is the following (with only a plausibility argument):

$$dx = dz,$$

$$f = f(x),$$

$$df = \frac{\partial f}{\partial x}\,dx + \frac{1}{2}\frac{\partial^2 f}{\partial x^2}\langle dx\,dx\rangle + \cdots,$$

$$df = \frac{\partial f}{\partial x}\,dx + \frac{1}{2}\frac{\partial^2 f}{\partial x^2}\,dt.$$

The second-order term in the Taylor expansion is first order in dt. More generally, if the stochastic process has drift and the function has time t dependence, then

$$dx = \mu(x,t)\,dt + \sigma(x,t)\,dz,$$

$$f = f(x,t),$$

$$df = \frac{\partial f}{\partial t}\,dt + \frac{\partial f}{\partial x}(\mu\,dt + \sigma\,dz) + \frac{1}{2}\frac{\partial^2 f}{\partial x^2}\sigma^2\,dt,$$

$$= \left(\frac{\partial f}{\partial t} + \mu\frac{\partial f}{\partial x} + \frac{\sigma^2}{2}\frac{\partial^2 f}{\partial x^2}\right)dt + \left(\sigma\frac{\partial f}{\partial x}\right)dz.$$

This is Ito's lemma; that is, there is a second-order derivative term in the drift of the process for the function in addition to the usual terms. This enables us to perform variable changes on the standard Brownian motion to obtain many more complex processes.

To start,

$$dx = \mu(x, t)\, dt + \sigma(x, t)\, dz(t).$$

Consider $S = e^x$. The process for S is given by

$$dS = S\left(\mu(\ln S, t) + \frac{\sigma(\ln S, t)^2}{2}\right) dt + S\sigma(\ln S, t)\, dz(t)$$

using Ito's lemma. This is a lognormal process for a constant σ value, which corresponds to the original case being normally distributed. We can make a slight change $\mu \to \mu - \frac{\sigma^2}{2}$ to obtain the general formula for the lognormal process

$$dS = S\mu\, dt + S\sigma\, dz(t).$$

Usually lognormal and many other continuous processes can all be related in this way. This means that we have described a large class of continuous processes (i.e., variance of steps is well-defined and goes to zero as time-step goes to zero), and they are not influenced by the past. Only the value at time t drives the future distribution. A process with this property is known as a Markovian process, that is, "time" correlation for these processes is zero. From the previous equation we get

$$\langle\, dz(t)\, dz(s)\rangle = dt\, ds\, \delta(t - s).$$

Thus we have built a formalism describing Markovian, continuous-time, stochastic processes (or Brownian motions). These will be the starting assumptions for many (but not all) models of security prices in finance.

For a general process for a security, we may write it as

$$dS = S\mu(S, t)\, dt + S\sigma(S, t)\, dz(t).$$

To relate this to reality, we note that the distribution—specifically means and variances of the path end-points for this more general process—are hard to calculate or estimate. The definition of *volatility* for a path looks at percentage changes, or returns on the security prices from sample time to sample time. Given an $N + 1$ time series of values for the security S, $(S_1, S_2, \ldots, S_{N+1})$ say, this generates N samples of returns, and a sample of variances of returns

$$r_i = \ln S_i,$$

$$\Delta r_i = \ln S_{i+1} - \ln S_i = \ln\left(\frac{S_{i+1}}{S_i}\right),$$

$$\text{Mean} = \langle \Delta r \rangle = \frac{1}{N} \sum_{i=1}^{N} \Delta r_i = \frac{1}{N} \sum_{i=1}^{N} (r_{i+1} - r_i),$$

$$\text{Variance} = \langle (\Delta r - \langle \Delta r \rangle)^2 \rangle = \frac{1}{(N-1)} \sum_{i=1}^{N} (\Delta r_i - \langle \Delta r \rangle)^2,$$

$$\text{Volatility} = \sqrt{\text{variance}}.$$

The formulae are the same as above; just differing by a variable change.

Again, clearly as $N \to \infty$, and the time-step goes to zero (and for a particular security price, S, and particular time, t), we would expect to get

$$\text{Volatility} = \sigma.$$

Again note that in reality it is going to be difficult to measure the volatility if there is time dependence of the term and very difficult if there is path value dependence. It would be necessary to perform multiple measurements of path value changes at nearly the same time and at very similar values for the path.

Looking more closely at Ito's lemma; there is an important relationship between Ito's lemma and the expectation of an exponential of a Gaussian distributed variable. From above (see section 2.5 Random Walks), we noted that for a Gaussian variable x_T with mean μT and standard deviation $\sigma \sqrt{T}$ (equivalently drift μ and volatility σ),

$$dx = \mu \, dt + \sigma \, dz,$$

$$x_T = \int_{t=0}^{t=T} dx,$$

$$\langle \exp(x_T) \rangle = \exp\left(\mu T + \frac{\sigma^2 T}{2} \right).$$

Now consider a lognormal process for stock price S, which at time T has value S_T,

$$dS = S\mu \, dt + S\sigma \, dz,$$

$$S_T = \int_{t=0}^{t=T} dS.$$

We want to find the value of S_T. More precisely, we want to find its expectation value and higher moments and so on. Consider a variable change

$S_T = e^{x_T}$ or $\ln S_T = x_T$, and then the process for x (using Ito's lemma) is

$$dx = \left(\mu - \frac{\sigma^2}{2}\right) dt + \sigma \, dz,$$

$$x_T = \int_{t=0}^{t=T} dx.$$

We note that the drift of x is slightly shifted from the drift term of S. We can more easily integrate this expression for the process of x (don't forget that the result has a distribution). And then, because S always has the value $S = e^x$, we know $S_T = e^{x_T}$. Then, using the expression for the expectation of the exponential of a Gaussian from above, we know

$$\langle S \rangle = \langle \exp(x_T) \rangle = \exp\left[\left(\mu - \frac{\sigma^2}{2}\right) T + \frac{\sigma^2}{2} T\right] = \exp(\mu T).$$

The point about this is that the extra term in the drift, due to Ito's lemma, cancelled out the term arising from taking the expectation of an exponential function of a random variable. Or, put another way, the Ito's lemma term arises to make sure that expectation values of functions of random variables are correct.

3.3. VARIABLE CHANGES TO GET THE MARTINGALE

The martingale is just the variable change under which a given process has zero drift. Consider

$$dx = \mu(x,t) \, dt + \sigma(x,t) \, dz,$$

$$f = f(x,t),$$

$$df = \frac{\partial f}{\partial t} dt + \frac{\partial f}{\partial x} (\mu \, dt + \sigma \, dz) + \frac{1}{2} \frac{\partial^2 f}{\partial x^2} \sigma^2 \, dt,$$

$$df = \left(\frac{\partial f}{\partial t} + \mu \frac{\partial f}{\partial x} + \frac{\sigma^2}{2} \frac{\partial^2 f}{\partial x^2}\right) dt + \left(\sigma \frac{\partial f}{\partial x}\right) dz.$$

Here, $f(x,t)$ is a martingale if

$$\frac{\partial f}{\partial t} + \mu \frac{\partial f}{\partial x} + \frac{\sigma^2}{2} \frac{\partial^2 f}{\partial x^2} = 0.$$

For example, if drift and volatility are pure constants (not time or security price dependent) then

$$f = e^x \exp\left(-\left[\mu + \frac{\sigma^2}{2}\right]t\right).$$

The meaning of this will become more apparent once we apply the formalism to finance (indeed, f turns out to be the futures price as opposed to the spot price $S = e^x$). For example, for options of strike K (in some currency, USD say), option maturity T, stock spot price S (in the same currency as strike), risk-less interest rate $r_\$$, and stock loan fee r_{stock}, a martingale at time t may be written as

$$m = \frac{Se^{-r_{stock}(T-t)}}{Ke^{-r_\$(T-t)}}$$

for the process

$$dS = S(r_\$ - r_{stock})\, dt + S\sigma(t)\, dz(t)$$

because the process for m

$$dm = m\sigma(t)\, dz(t)$$

has no drift. Generally if we have two variables each with its own process, say,

$$dS_1 = S_1(r_\$ - r_1)\, dt + S_1\sigma_1(t)dz_1(t),$$
$$dS_2 = S_1(r_\$ - r_2)\, dt + S_2\sigma_2(t)dz_2(t),$$

then a martingale can be constructed from the ratio (or indeed the product of any powers of the two variables) as m, and its process is given by

$$m = \frac{S_1}{S_2}e^{g(t)},$$

$$dm = m\left(\frac{dS_1}{S_1} - \frac{dS_2}{S_2} + (g^2(t) + \sigma_2^2(t))\, dt\right),$$

$$= m\left(\sigma_1(t)dz_1(t) - \sigma_2(t)dz_2(t) + \left(r_2 - r_1 + \frac{\partial g(t)}{\partial t} + \sigma_2^2(t)\right)dt\right),$$

which implies that m is a martingale if we choose

$$g(t) = -\int_0^t [r_2(s) - r_1(s) + \sigma_2^2(s)]\, ds.$$

3.4. OTHER PROCESSES: MULTIVARIABLE CORRELATIONS

Given two independent Wiener processes $dz_1(t)$ and $dz_2(t)$, we can construct two arbitrary paths $X_1(t)$ and $X_2(t)$, of different means and standard deviations and arbitrary correlation of changes. (The analysis has already been done in section 2.6 Correlation.)

$$dx_1 = \alpha_1(t)(dz_1(t) + \beta(t)dz_2(t)) + \mu_1(t)\, dt,$$

$$dx_2 = \alpha_2(t)(dz_1(t) - \beta(t)dz_2(t)) + \mu_2(t)\, dt;$$

$$dz_1 = \frac{1}{2}\left[\frac{(dx_1 - \mu_1\, dt)}{\alpha_1} + \frac{(dx_2 - \mu_2\, dt)}{\alpha_2}\right],$$

$$dz_2 = \frac{1}{2\beta}\left[\frac{(dx_1 - \mu_1\, dt)}{\alpha_1} - \frac{(dx_2 - \mu_2\, dt)}{\alpha_2}\right];$$

$$\alpha_1(t) = \sigma_1(t)\sqrt{\frac{1+\rho(t)}{2}}, \quad \alpha_2(t) = \sigma_2(t)\sqrt{\frac{1+\rho(t)}{2}}, \quad \beta(t) = \sqrt{\frac{1-\rho(t)}{1+\rho(t)}}.$$

Note that these changes are normally distributed (not lognormally; we have a potential model of stock price returns):

$$\langle dx_1 \rangle = \mu_1(t)\, dt, \quad \langle (dx_1 - \langle dx_1 \rangle)^2 \rangle = \sigma_1^2(t)\, dt;$$

$$\langle dx_2 \rangle = \mu_2(t)\, dt, \quad \langle (dx_2 - \langle dx_2 \rangle)^2 \rangle = \sigma_2^2(t)\, dt;$$

$$\frac{\langle dx_1 dx_2 \rangle - \langle dx_1 \rangle \langle dx_2 \rangle}{\sigma_1(t)\sigma_2(t)} = \rho\, dt.$$

The path end-points are given by

$$X_1 = \int_{t=0}^{t=T} dx_1,$$

$$X_2 = \int_{t=0}^{t=T} dx_2$$

and they have a distribution given by (from section 2.6 Correlation)

$$P(X_1, X_2)dX_1 dX_2 = \frac{dX_1 dX_2}{2\pi \Sigma_1 \Sigma_2 T\sqrt{1-P^2}} \exp\left[-\frac{1}{2(1-P^2)}\left\{\left(\frac{X_1 - M_1 T}{\Sigma_1\sqrt{T}}\right)^2 \right.\right.$$

$$\left.\left. + \left(\frac{X_2 - M_2 T}{\Sigma_2\sqrt{T}}\right)^2 - 2P\left(\frac{X_1 - M_1 T}{\Sigma_1\sqrt{T}}\right)\left(\frac{X_2 - M_2 T}{\Sigma_2\sqrt{T}}\right)\right\}\right].$$

The means, standard deviations and correlation of the path endpoints (expressed as averages here) are given by

$$M_1 = \frac{1}{T} \int_{t=0}^{t=T} \mu_1(t)\, dt,$$

$$M_2 = \frac{1}{T} \int_{t=0}^{t=T} \mu_2(t)\, dt,$$

$$(\Sigma_1)^2 = \frac{1}{T} \int_{t=0}^{t=T} \sigma_1^2(t)\, dt,$$

$$(\Sigma_2)^2 = \frac{1}{T} \int_{t=0}^{t=T} \sigma_2^2(t)\, dt,$$

$$P = \frac{1}{T} \frac{\int_{t=0}^{t=T} \sigma_1(t)\sigma_2(t)\rho(t)\, dt}{\Sigma_1 \Sigma_2}.$$

Note that the average correlation P does not grow in the distribution. It does not have a factor T next to it in the distribution of the path endpoints. Also note that the simplification for means, volatilities, and correlation are time independent:

$$M_1 = \mu_1, M_2 = \mu_2, \Sigma_1 = \sigma_1, \Sigma_2 = \sigma_2, P = \rho.$$

In summary then, we have related the means, standard deviations and correlations of two paths' endpoints to the drift rates, volatilities, and correlation of the two paths' changes. Furthermore, and most importantly for practical modeling, we have related the quite complicated analysis of path distributions of variables with arbitrary functions of time for drift, volatilities and correlation to the very simple distribution of two standard deviation 1, mean 0, uncorrelated Gaussian distributed variables.

CHAPTER 4

Applications of Stochastic Calculus to Finance

4.1. RISK PREMIUM DERIVATION

The following is a very surprising result. For a single Wiener process driving a market, there is a universal *risk premium* (or price of risk). Consider two derivatives of this driver that trade in this market,

$$dS_1 = S_1 \mu_1 \, dt + S_1 \sigma_1 \, dz(t),$$

$$dS_2 = S_2 \mu_2 \, dt + S_2 \sigma_2 \, dz(t),$$

where the drift and volatilities may be arbitrary functions of time and their respective stocks. We may think of security 1 as an option and security 2 as a stock. Note that they are perfectly correlated because the Wiener process is the same for each security (admittedly, this is not a very realistic model of a market). There may be an infinite number of different securities from which to choose the two securities.

Construct a portfolio that is *long* one security and *short* Δ shares of the other:

$$\Pi = S_1 - \Delta S_2.$$

LONG AND SHORT

Long is a finance term meaning to own an amount. *Short* means to effectively own a negative amount. The latter is realized in the capital markets by borrowing a fungible security and immediately selling it—and then buying it back at a later date and returning this under the borrow agreement. *Fungible* means securities for which this transaction is allowed.

Now consider the process for this portfolio:

$$d\Pi = dS_1 - \Delta dS_2$$
$$= (\mu_1 S_1 - \Delta \mu_2 S_2)\, dt + (\sigma_1 S_1 - \Delta \sigma_2 S_2)\, dz(t).$$

Carefully note that the number of shares of security 2 that we are short—that is, Δ—may be chosen as a function of time and stock price to ensure that the portfolio is risk-free. Its values are stochastic (i.e. have a distribution), but this is only because the stock prices themselves are stochastic. We then choose Δ so that the portfolio has only drift and no stochastic process. Indeed, if we choose

$$\Delta = \frac{\sigma_1 S_1}{\sigma_2 S_2},$$

then the portfolio has zero risk, in the sense that there is no variance. The portfolio is now synthetic cash. We will earn risk-free interest on the cash value of the portfolio plus earning the value of stock borrow on stock 2 (because we have to pay this to finance the short stock 2 position) minus stock loan on stock 1 (because we could lend it out now that we own it):

$$d\Pi = (\mu_1 S_1 - \Delta \mu_2 S_2)\, dt = (-r_{stock1} S_1 + r_\$ \Pi + \Delta S_2 r_{stock2})\, dt$$

Rearranged we get

$$\frac{(\mu_1 - r_\$ + r_{stock1})}{\sigma_1} = \frac{(\mu_2 - r_\$ + r_{stock2})}{\sigma_2}.$$

Now, because there are infinitely many different derivatives of the Wiener process driving this market, this relationship must hold for any two and therefore all securities. Thus, only solution is quite remarkable:

$$\frac{(\mu_1 - r_\$ + r_{stock1})}{\sigma_1} = \lambda.$$

There must be a universal constant λ, (possibly a function of t, possibly even stochastic itself) governing the excess return for each security over the risk-free financing costs of a position in that security:

$$\mu_1 = r_\$ - r_{stock1} + \lambda \sigma_1.$$

Each security's excess return (or drift/growth rate in excess of risk-less financing costs) must be proportional to the volatility, or risk, of the specific security. This proportionality factor is a universal market constant; the risk premium.

4.2. ANALYTIC FORMULA FOR THE EXPECTED PAYOFF OF A EUROPEAN OPTION

Assuming a lognormal stock price process, we can now do some very valuable calculations. The expected value of the payoff is an example. For a European equity call option that expires at time T, the payoff value (in dollars) is a function $C_T = \max(S_T - K, 0)$ and the question is: What is the expected value today $(t = 0)$ of this payoff, say $E(payoff)$? Clearly, if we have a terminal probability distribution over final stock prices S_T, that is, $P(S_T, S)dS_T$, given that we start at stock price S, then the answer is the weighted sum of terminal probabilities times payoffs added up over all terminal stock prices (i.e., a convolution of two functions), which is expressed as

$$E(payoff) = e^{-rsT} \int_{S_T=0}^{S_T=\infty} P(S_T, S)C_T(S_T)dS_T.$$

A discount factor is added to ensure the result is a dollar value today (rather than a dollar value at maturity). Note that this is a candidate for the price of the security; but we will soon see that it is *not* the price. Nevertheless, let's proceed.

The stock price process we assume to be

$$dS_t = S_t\mu(t)\, dt + S_t\sigma(t)\, dz(t),$$

and then change variables to $x_t = \ln(S_t/S)$, where S is the *initial* value of the stock-price path at time $t = 0$ and $x = 0$. S_t is the stock price at an intermediate time t, (i.e., $0 \leq t \leq T$) and similarly for x_t. Then the process may be rewritten using Ito's lemma as

$$dx_t = \left(\mu(t) - \frac{\sigma(t)^2}{2} \right) dt + \sigma(t)\, dz(t).$$

We know this Gaussian process has terminal distribution

$$P(x_T)dx_T = \frac{1}{\sqrt{2\pi \Sigma^2 T}} \exp\left[-\frac{(x_T - MT)^2}{2\Sigma^2 T} \right] dx_T$$

where

$$M = \frac{1}{T} \int_{t=0}^{t=T} \left(\mu(t) - \frac{\sigma(t)^2}{2} \right) dt,$$

$$\Sigma^2 = \frac{1}{T} \int_{t=0}^{t=T} \sigma(t)^2 dt.$$

Let's make a further variable change:

$$z = \frac{(x_T - MT)}{\Sigma \sqrt{T}},$$

$$P(x_T)dx_T = P(z) dz = \frac{1}{\sqrt{2\pi}} \exp \left[-\frac{z^2}{2} \right] dz,$$

$$C_T(z) = \max(S \exp(z\Sigma\sqrt{T} + MT) - K, 0).$$

So the answer (in terms of integration over the z variable) is

$$E(payoff) = \exp(-r_\$T) \int_{z=-\infty}^{\infty} C_T(z)P(z) dz$$

$$= \exp(-r_\$T) \int_{z=z_K}^{\infty} P(z)[S \exp(z\Sigma\sqrt{T} + MT) - K] dz$$

where

$$z_K = \frac{\ln \left(\frac{K}{S} \right) - MT}{\Sigma \sqrt{T}}.$$

Cumulative norm functions are involved—and thus no analytic result is available without integrals—so if we use the definition of a cumulative norm function,

$$N(x) = \frac{1}{\sqrt{2\pi}} \int_{z=-\infty}^{z=x} \exp \left(-\frac{z^2}{2} \right) dz,$$

then we can write the solution as

$$E(payoff) = \exp(-r_\$T) \int_{z=z_K}^{\infty} P(z)[S \exp(z\Sigma\sqrt{T} + MT) - K] dz$$

$$= Se^{\left(\left[M + \frac{\Sigma^2}{2} - r_\$ \right] T \right)} N(d_+) - Ke^{(-r_\$T)} N(d_-)$$

where

$$d_+ = \frac{\ln\left(\frac{S}{K}\right) + MT + \Sigma^2 T}{\Sigma\sqrt{T}},$$

$$d_- = \frac{\ln\left(\frac{S}{K}\right) + MT}{\Sigma\sqrt{T}}.$$

One last step is needed. We rewrite the expected payoff in terms of the average drift μ_0, and average annualized volatility, σ_0, as

$$\mu_0 = \frac{1}{T}\int_{t=0}^{t=T} \mu(t)\,dt,$$

$$\sigma_0 = \sqrt{\Sigma^2}.$$

We then get, using the average continuous discount rate ("instantaneous forward"),

$$E(payoff) = Se^{([\mu_0 - r_\$]T)}N(d_+) - Ke^{(-r_\$ T)}N(d_-),$$

$$d_\pm = \frac{\ln(S/K) + \mu_0 \pm \frac{1}{2}\sigma_0^2 T}{\sigma_0\sqrt{T}}.$$

Noting from above (see section 4.1 Risk Premium Derivation), that

$$\mu_0 = r_\$ - r_S + \lambda\sigma_0$$

and substituting this in we get

$$E(call\ payoff) = Se^{([-r_S + \lambda\sigma_0]T)}N(d_+) - Ke^{(-r_\$ T)}N(d_-),$$

$$d_\pm = \frac{\ln\left(\frac{Se^{([-r_S + \lambda\sigma_0]T)}}{Ke^{(-r_\$ T)}}\right) \pm \frac{1}{2}\sigma_0^2 T}{\sigma_0\sqrt{T}}.$$

Also note that a put option has an expectation payoff value similarly expressed as

$$E(put\ payoff) = -Se^{([-r_S + \lambda\sigma_0]T)}N(-d_+) + Ke^{(-r_\$ T)}N(-d_-).$$

This *is* the correct value for the "expectation" payoff of the options.

However, the fair-price is (perhaps surprisingly) *different*. We will see in a subsequent chapter that the fair price (or arbitrage-free price) comes out to be the same expressions, for call and put, but with the value of zero for the risk premium, $\lambda = 0$. We would have obtained the correct answer if we had assumed this value for the risk premium or, in other words, valued the expectation payoff of the option in a risk-neutral world. The difference between the two values for the option, that is,

$$E(payoff, \lambda) - E(payoff, \lambda = 0)$$

is the present value of a particular dynamic hedging strategy that reduces the market risk, i.e., variance, of the portfolio consisting of option plus hedging strategy, to zero. (This concept is discussed in detail later in this book.)

Before discussing risk-neutral pricing—and pricing a call and put correctly—we want to introduce the idea of a differential equation that the terminal probability density solves. That is, an *equation of motion*.

CHAPTER 5

From Stochastic Processes Formalism to Differential Equation Formalism

5.1. BACKWARD AND FORWARD KOLMOGOROV EQUATIONS

In this section no results will be derived per se. Some results, however, relevant to our topic will be stated.

Given a generalized price process,

$$dS = \mu(S, t)\, dt + \sigma(S, t)\, dz(t),$$

$$S_T = S_t + \int_{t'=t}^{t'=T} dS(t'),$$

with arbitrary drift function and local volatility function, we may ask the following questions:

- Given that we start at stock price S_t at time t, what is the probability distribution over S_T at some time $T > t$?
- What are the ending expectation values of observables (i.e., expectation values of functions of the path end-point or state variable)?
- What are the expectation values of these observables as functions of the path start-points S_t ?

It turns out that a probability distribution exists $P(S_t, S_T)$, which encapsulates all this and satisfies two equations (of motion) in the two sets of variables. The first is called the *Fokker-Planck* (also called the *forward Kolmogorov*) equation:

$$\left(-\frac{\partial}{\partial T} + \frac{\partial^2}{\partial S_T^2} \frac{\sigma(S_T, T)^2}{2} - \frac{\partial}{\partial S_T} \mu(S_T, t) \right) P(S_t, S_T, T, t) = 0.$$

(Note: As a reminder, this *operator notation* implies that the derivative operators act on everything to the right.) The second is the *backward Kolmogorov equation:*

$$\left(\frac{\partial}{\partial t} + \frac{\sigma(S_t, t)^2}{2}\frac{\partial^2}{\partial S_t^2} + \mu(S_t, t)\frac{\partial}{\partial S_t}\right) P(S_t, S_T, T, t) = 0.$$

For example, for a constant mean and standard deviation μ_0, σ_0, that is, a pure Gaussian process

$$dx = \mu_0\, dt + \sigma_0\, dz(t),$$

$$x_T = x_t + \int_{t'=t}^{t'=T} dx(t') :$$

the two equations, which the probability distribution of the path solves, look more alike (just a remaining minus sign difference):

$$\left(-\frac{\partial}{\partial T} + \frac{\sigma_0^2}{2}\frac{\partial^2}{\partial x_T^2} - \mu_0\frac{\partial}{\partial x_T}\right) P(x_t, x_T, T, t) = 0,$$

$$\left(\frac{\partial}{\partial t} + \frac{\sigma_0^2}{2}\frac{\partial^2}{\partial x_t^2} + \mu_0\frac{\partial}{\partial x_t}\right) P(x_t, x_T, T, t) = 0,$$

and the solution (to both equations) is

$$P(x_t, t; x_T, T) = \frac{1}{\sqrt{2\pi\sigma_0^2(T-t)}} \exp\left(-\frac{((x_T - x_t) - \mu_0(T-t))^2}{2\sigma_0^2(T-t)}\right).$$

$P(x_t, t; x_T, T)$ is a very powerful function. We may view it as a transition probability between the two states $\langle x_t, t|$ and $|x_T, T\rangle$. The first (left) state is a starting state and the latter (right) state is a finishing state. The transition probability is the "dot" product (or inner product, in the sense of vectors, albeit infinite dimensional vectors) between them:

$$P(x_t, t; x_T, T) = \langle x_t, t|x_T, T\rangle.$$

Furthermore, expectation values of functions of the state variables $f(x_t, t) = \langle f|x_t, t\rangle$ can be calculated at any later time,

$$f(x_T, T) = \int_{x_t=-\infty}^{x_t=+\infty} f(x_t, t) P(x_t, t; x_T, T)\, dx_t$$

$$= \int_{x_t=-\infty}^{x_t=+\infty} \langle f|x_t, t\rangle\langle x_t, t|x_T, T\rangle\, dx_t = \langle f|x_T, T\rangle,$$

or, if $f(x_T, T) = \langle x_T, T | f \rangle$ is given, for an earlier time,

$$
\begin{aligned}
f(x_t, t) &= \int_{x_T=-\infty}^{x_T=+\infty} f(x_T, T) P(x_t, t; x_T, T)\, dx_T \\
&= \int_{x_T=-\infty}^{x_T=+\infty} \langle x_t, t | x_T, T \rangle \langle x_T, T | f \rangle\, dx_T = \langle x_t, t | f \rangle,
\end{aligned}
$$

depending on whether the function f has its values defined over the time slice t or T, respectively. For example, an expectation payout (see section 4.2 Analytic Formula for the Expected Payoff of a European Option) is an example representing an observable value that is defined on the time slice T, and then has an expectation value at time $t = 0$, defined over stock price S.

Note the analysis in this section is without the time value of money which means that the discounting due to the time value of the dollar is removed, and also T is replaced by $T - t$ from the expectation value formulae. Therefore, the call expectation value expression is

$$
E(call\ payoff, S, t;\ T) = Se^{(r_S - r_S + \lambda\sigma_0)(T-t)} N(d_+) - K N(d_-),
$$

which must solve the backward Kolmogorov equation,

$$
\left(\frac{\partial}{\partial t} + \frac{\sigma_0^2 S^2}{2} \frac{\partial^2}{\partial S^2} + S\mu_0 \frac{\partial}{\partial S} \right) E(call\ payoff, S, t;\ T) = 0
$$

for the stock price process, $dS = S\mu_0\, dt + S\sigma_0\, dz(t)$.

The Kolmogorov equations are a powerful tool because, in situations that cannot be solved analytically, we can always find a numerical solution to the differential equation to get expectation values with the input of a boundary condition. In the case of a call or put option expectation value, this boundary condition is the terminal option value at $t = T$,

$$
C_T = \max(S_T - K, 0).
$$

We then just (numerically if necessary) integrate the backward Kolmogorov equation (backwards in time) to time t, to get the expectation value.

Finally, observe that the backward Kolmogorov equation is just the heat equation if it is rewritten in a spatial variable x such that the coefficient of the second-order (curvature) term has no functional dependence on x. (A variable transform will find these *normal coordinates*.) Thus

$$\left(\frac{\partial}{\partial t} + \frac{\sigma_0^2}{2}\frac{\partial^2}{\partial x^2} + \mu(x,t)\frac{\partial}{\partial x}\right)\Theta(x_t, x_T, T, t) = 0,$$

where Θ is temperature (i.e., expectation value), and if the temperature of the bar is described as observed from a car moving at velocity

$$\mu(x,t)$$

with time running backwards! If we started out with an infinitely long bar and temperature Θ where $\Theta = 0$ for $x < 0$ and $\Theta = \exp(x) - 1$ for $x > 0$, then the solution is the same as the expectation values for the payoffs of the put and call as above (see section 4.2 Analytic Formula for the Expected Payoff of a European Option). It would gradually heat up around the temperature "kink" (the strike), because heat transfers from the high-temperature to the low-temperature area, while the long arm of temperature gradient out at large x appears to gradually cool down; this is, in fact, just due to the translation. Backwards in time for options; the heating due to heat transfer is option premium decay; and the translation is stock forward compounding.

5.2. DERIVATION OF BLACK-SCHOLES EQUATION, RISK-NEUTRAL PRICING

We now consider an evaluation of the fair price of a derivative. We do it by constructing a portfolio of derivative plus hedging strategy that results in a portfolio value that has no variance at any time (i.e., it is riskless) and then the fair value of this portfolio is evident because it is "synthetic riskless cash" and so it accretes at the riskless rate.

At time t, the portfolio contains cash; an option (or indeed any derivative) of value $C(S,t)$; and a stock of value S and of quantity $-\Delta$; the borrow costs for the stock and cash are r_S, $r_\$$ and, although it is almost always ignored, I shall include a term for the option borrow cost as r_C.

Now construct a portfolio, of value Π at time t, as follows: borrow cash in the amount $C - \Delta S$ generating a liability; borrow Δ shares of stock, generating a stock liability; and purchase one option. It is expressed as

$$\Pi = \Pi_\$ + \Pi_S + \Pi_C,$$
$$\Pi_\$ = -(C - \Delta S),$$
$$\Pi_S = -\Delta S,$$
$$\Pi_C = C.$$

We have three products in the portfolio: dollars, stock and one option. They are all written in terms of their values in \$. At time t the total portfolio value is zero. Then consider the portfolio value at time $t + dt$. If the stock price process is

$$dS = S\mu dt + S\sigma\, dz(t),$$

then we are able to derive the process for the portfolio value (using Ito's lemma when necessary), that is

$$d\Pi = d\Pi_\$ + d\Pi_S + d\Pi_C,$$

$$d\Pi_\$ = -(C - \Delta S)r_\$ \, dt - \Delta S r_S \, dt + C r_C \, dt,$$

$$d\Pi_S = -\Delta(S\mu dt + S\sigma\, dz(t)),$$

$$d\Pi_C = \left(\frac{\partial C}{\partial t} + \frac{S^2\sigma^2}{2}\frac{\partial^2 C}{\partial S^2} + S\mu\frac{\partial C}{\partial S}\right) dt + S\sigma\frac{\partial C}{\partial S}\, dz(t).$$

Carefully note that the cash subportfolio generates more liability from financing the cash we already borrowed, income from lending out the option we own, and extra liability paying the borrow cost on the stock we borrowed. If at time t we choose to short enough shares to make the variance of the whole portfolio zero (i.e., make the coefficient of the $dz(t)$ term zero), then

$$\Delta = \frac{\partial C}{\partial S}$$

and we know that its total value at time $t + dt$ must be zero, as it was at time t. Mathematically, the only remaining term is the drift term, and this must also be zero, for this riskless portfolio of value zero:

$$\frac{\partial C}{\partial t} + \frac{S^2\sigma^2}{2}\frac{\partial^2 C}{\partial S^2} + S(\mu - \mu + r_\$ - r_S)\frac{\partial C}{\partial S} - (r_\$ - r_C)C = 0.$$

The interpretation of all the "interest" rates $r_\$, r_S, r_C$ is that they are all riskless rates, because this portfolio was riskless.

Usually we are looking for the "arbitrageable" value of the option (i.e., the riskless value we can earn from it), and so we ignore the r_C term because it is just this "yield-to-maturity" that we are trying to evaluate, that is, the option cheapness in terms of arbitraging the option price (or implied volatility) versus the underlying stock volatility. This leaves us with the derivative pricing equation—that is, the Black-Scholes equation:

$$\frac{\partial C}{\partial t} + \frac{S^2\sigma^2}{2}\frac{\partial^2 C}{\partial S^2} + S(r_\$ - r_S)\frac{\partial C}{\partial S} - r_\$ C = 0.$$

The financial interpretation of this equation is as follows: rewriting slightly,

$$\Delta C = \frac{\partial C}{\partial t}\Delta t = -\frac{S^2\sigma^2}{2}\frac{\partial^2 C}{\partial S^2}\Delta t + r_\$\Delta t(C - \Delta S) + r_S\Delta t S\Delta,$$

the change in option price, or theta, is equal to the sum of the negative of the profits generated from delta re-hedging (let's call this "decay"), plus the carry on the cash in the position (to finance the net long cash value), plus the carry on the short stock (to finance the stock borrow). The right-hand side is the negative of the "riskless portfolio" that replicates an option price integrated over time, and so the option plus this value is zero.

5.3. RISKS AND TRADING STRATEGIES

The previous section outlines a theoretical method of trading options that, under the assumptions outlined (stock price changes that are continuous Markovian processes and no credit or default risks whatsoever), reduces the remaining "risk" (defined as variance of returns) to zero.

More generally the idea is that, even in the real world, certain dynamic hedging strategies may reduce risks (although never sending all risks to zero unless you sell all positions completely). For instance, we may use the Black-Scholes equation to give us a dynamic hedging strategy for an option using stock trading. This is buying and selling stock according to the delta,

$$\Delta = \frac{\partial C}{\partial S}$$

of the option position. But we still own something. This is because we have a "long position" (i.e., a positive amount) of the option implied volatility and a "short position" (i.e., an effectively negative amount) of the realized volatility of the underlying stock. This method of trading will realize any difference between these two numbers. This realized profit and loss (P&L) can be measured by *vega* (approximately, because now we are assuming the Black-Scholes assumptions do not hold), that is

$$vega = \frac{\partial C}{\partial \sigma}$$

and its interpretation is as follows:

If realized stock volatility is measured using the stock prices at which delta rehedge trades are executed over the entire life of the option, then vega is the profit made (or lost) for each point that realized stock volatility is above (or below) the implied volatility.

Vega is also the instantaneous P&L for a change in implied volatility of the option.

Similarly, the interest rate risk may be measured and dynamically hedged

$$\rho = \frac{\partial C}{\partial r_\$}.$$

This number ρ, measures the instantaneous risk of one option for a parallel shift in the interest rate curve. Another risk measure for dividend risk (or indeed stock borrow risk) is $\rho_1 - \frac{\partial C}{\partial r_{Stock}}$. Further risk measures are possible for yield curve shape changes and so on, although for a European option (exercisable only at maturity), the only interest rate risk is the "today to expiration" forward rate or maturity zero price. All other zero prices are independent. (For an argument for why this is so, see Chapter 7 Interest Rate Hedging.)

By trading pure interest rate derivatives, we can generate a dynamic hedging strategy to significantly reduce interest rate risk. We could even trade other equity derivatives to mitigate the vega risk as well. All this using a model that does *not* assume stochastic volatility or rates! How is this possible? Generally the argument is as follows: The Black-Scholes derivation in fact only assumed a lognormal distribution of changes of the martingale

$$\frac{Se^{(-r_\$T)}}{Ke^{(-r_\$T)}}$$

and this implies that a reinterpretation of the volatility will lead to similar numbers to a full stochastic rates model. Second, the main risk to an option is volatility and so the inclusion of models of the other components gives better answers—assuming they are correct—for ρ and vega. However, these corrections will be smaller than the initial *contribution* of including the main driver to an option value, the martingale volatility. Thus, we certainly would expect to decrease variance of returns by hedging the ρ and vega numbers, even if the Black-Scholes model is wrong or even if we get the inputs such as volatility slightly wrong.

The conclusion is that the Black-Scholes derivation (and risk-neutral pricing generally) is not actually about risk-neutrality *per se* in its application. It is about which risks are reduced in order to leave accentuated *unhedged* risks that the trader has a view on and wants to own or arbitrage.

Therefore the basic option versus delta-traded stock strategy is a volatility arbitrage trade and requires the reduction of market (or delta) risk and a view on whether implied volatility is rich or cheap versus expected realized volatility. Or the trader could even short other derivatives that are market neutral and have vega, leaving the trade long a volatility spread (and neutral total vega). This may well be a dynamic strategy as well. The trader has taken a view on one price of volatility (i.e., implied volatility) versus another and is trading that spread while taking no view on the market's overall volatility.

Understanding the Black-Scholes Equation

T he Black-Scholes equation has many remarkable features. The important ones are outlined and discussed in this chapter.

6.1. BLACK-SCHOLES EQUATION: A TYPE OF BACKWARD KOLMOGOROV EQUATION

First, the final term in the Black-Scholes equation is the discounting due to the time value of cash; if we wrote the formula in terms of the function $C'(S, t)$ as

$$C(S, t) = \exp\left(-\int_{s=t}^{s=T} r_\$(s)\, ds\right) C'(S, t),$$

which is the option value in maturity dollars, then the last term would disappear as

$$\frac{\partial C'}{\partial t} + \frac{S^2\sigma^2}{2}\frac{\partial^2 C'}{\partial S^2} + S(r_\$ - r_S)\frac{\partial C'}{\partial S} = 0.$$

It is immediately evident that it is a type of backward Kolmogorov equation—but with the "wrong" drift term for the stock price process because the drift of stock includes the risk premium in reality. We have already solved this problem for drift and volatility as arbitrary functions of time but *not* functions of stock price. So, for this case, from section 4.2 Analytic Formula for the Expected Payoff of a European Option, we can immediately write the Black-Scholes option *pricing* formulae as

$$C(S, t) = E_{risk-neutral}(call\ payoff) = Se^{(-r_S T)}N(d_+) - Ke^{(-r_\$ T)}N(d_-),$$

$$C(S, t) = E_{risk-neutral}(put\ payoff) = -Se^{(-r_S T)}N(-d_+) + Ke^{(-r_\$ T)}N(-d_-),$$

$$d_{\pm} = \frac{\ln\left(\frac{Se^{(-r_S T)}}{Ke^{(-r_\$ T)}}\right) \pm \frac{1}{2}\sigma^2 T}{\sigma\sqrt{T}},$$

where the expectation value is understood to be the risk-neutral expectation value, and volatilities and rates are to be understood as averages of their possibly term-structured values (i.e., functions of time) taken over the life of the options. Thus

$$r_\$ = \frac{1}{T}\int_{t=0}^{t=T} r_\$(t)\,dt,$$

$$\sigma^2 = \frac{1}{T}\int_{t=0}^{t=T} \sigma(t)^2\,dt.$$

Note the following alternative solutions to the Black-Scholes equation,

$$f_0 = K\exp(-r_\$(T-t)),$$
$$f_1 = S\exp(-r_S(T-t)),$$
$$f_2 = S^2\exp(-r_S(T-t))\exp((\sigma^2 + 2(r_\$ - r_S))(T-t)),$$

which represent the first three moments of the (terminal) stock price distribution: the "zero-th" moment is the normalization, which is cash discounting (to get a zero-coupon-bond price correct), then the first moment is the stock forward price, always a discount to the current stock. This lines up the forwards or calibrates the model in terms of the prices of the two products embedded in the option. The second moment of the terminal distribution contains the volatility (or standard deviation, i.e., second moment, of the stock price changes). The equivalent formulae for the first three moments that solve the backward Kolmogorov equation, that is, just the above moments divided by $\exp(-r_\$(T-t))$, are as follows:

$$g_0 = 1,$$
$$g_1 = S\exp((r_\$ - r_S)(T-t)) = \langle S_T\rangle|_{S_0=S},$$
$$g_2 = S^2\exp((\sigma^2 + 2(r_\$ - r_S))(T-t)) = \langle S_T^2\rangle|_{S_0^2=S^2}.$$

The first term is the normalization of the distribution, the second term is the first moment, and it shows that the stock price grows to match up with the "forward price" under risk-neutral pricing, while the third term shows the growth of the value of stock squared.

6.1.1. Forward Price

In finance, the *forward price* is essentially a price or an exchange rate, agreed upon today, at which to swap two products at a later date, the *forward date*. This price will clearly be different depending on risks and the like specified in the language of the forward contract. It will also, however, be different simply because of the discounting difference between the two products being exchanged.

Thus the volatility measured from time t to T, has expectation value

$$\text{Volatility}^2 = \frac{1}{(T-t)} \frac{\text{variance}(S_T)}{\text{mean}(S_T)^2} = \frac{1}{(T-t)} \frac{\left[\langle S_T^2 \rangle |_{S_0^2 = s^2} - \left(\langle S_T \rangle |_{S_0 = s} \right)^2 \right]}{\left(\langle S_T \rangle |_{S_0 = s} \right)^2}$$

$$= \frac{\exp(\sigma^2(T-t)) - 1}{(T-t)},$$

which relates measured volatility to instantaneous volatility, because the limit $(T - t) \to 0$ is given by

$$\lim_{(T-t) \to 0} \text{volatility}^2 = \sigma^2.$$

6.2. BLACK-SCHOLES EQUATION: RISK-NEUTRAL PRICING

Very importantly, the risk-premium λ dropped out during the derivation of the Black-Scholes equation. Compare to the backward Kolmogorov equation, which most certainly contains the full drift term μ including the risk premium λ. This drift independence, or pricing under a risk-neutral probability measure, is the central result of the work of Black, Scholes and Merton in the early 1970s. It is fair to say that this insight, that the general drift of the market does not affect option prices, fueled the growth of the option markets through the 1970s and 1980s. The expectation value formula had been known since the early 20th century, but nobody knew what drift to put in the formula. It turned out that the price is not the expectation value formula: it is the expectation value with the risk-premium set to zero. We can write

$$\text{Price} = E(payoff, \lambda = 0)$$

$$= E(payoff, \lambda) + [E(payoff, \lambda = 0) - E(payoff, \lambda)],$$

splitting the option price into the present value of the expectation payoff, and another term. The price has zero variance, but the option expectation value has a variance. The second term is the present value (PV) of the dynamic hedging strategy; shorting

$$\Delta = \frac{\partial C}{\partial S}$$

shares at all times (i.e., rehedging every instant). The variance of this term exactly cancels the variance of the expectation value to give a (zero variance, arbitrage-free) price. The price turns out to be the expectation value of the payoff under a risk-neutral pricing scenario.

6.3. BLACK-SCHOLES EQUATION: RELATION TO RISK PREMIUM DEFINITION

Next we note the relation to the previous derivation of the definition of the market risk premium λ defined by

$$\mu_S = r_\$ - r_S + \lambda\sigma_S,$$
$$\mu_C = r_\$ - r_C + \lambda\sigma_C,$$

rewrite the second equation as

$$\mu_C C - (r_\$ - r_C + \lambda\sigma_C)C = 0,$$

and, noting that the processes for C and S are

$$dS = S\mu_S \, dt + S\sigma_S \, dz(t),$$
$$dC = C\mu_C \, dt + C\sigma_C \, dz(t),$$

use Ito's lemma

$$C\mu_C = \frac{\partial C}{\partial t} + \frac{S^2\sigma_S^2}{2}\frac{\partial^2 C}{\partial S^2} + S\mu_S\frac{\partial C}{\partial S},$$
$$C\sigma_C = S\sigma_S\frac{\partial C}{\partial S},$$

to show that the risk-neutral definition equation for the option is

$$\mu_C C - (r_\$ - r_C + \lambda\sigma_C)C = 0,$$
$$\frac{\partial C}{\partial t} + \frac{S^2\sigma_S^2}{2}\frac{\partial^2 C}{\partial S^2} + S(\mu_S - \lambda\sigma_S)\frac{\partial C}{\partial S} - (r_\$ - r_C)C = 0.$$

The risk premium has dropped out. The Black-Scholes equation *is* the risk premium definition equation in the sense that imposing the Black-Scholes equation on the prices of derivatives ensures that the processes those derivative's prices satisfy have an average instantaneous growth equal to the risk-less interest rate plus the market risk premium times the derivative's volatility.

6.4. BLACK-SCHOLES EQUATION APPLIES TO CURRENCY OPTIONS: HIDDEN SYMMETRY 1

The formalism here applies to currency options. Interpret the stock price S as the exchange rate to another currency, say euros; more precisely S is the value of 1 euro, in $ (instead of the value of one share in $). Then, the right but not the obligation to buy 1 euro at some exchange rate K, that is, exchange $K$$ for 1 euro, is known as a call on euros. The payoff is the same as for a stock option.

However, this is the same security as a put on $, priced in euros (more precisely K puts). To see this, we show hidden symmetry 1 of the Black-Scholes equation.

A call on 1 euro, for $K$$ is priced as

$$C(S, T) = \max(S - K, 0),$$

$$C(S, t): \quad \frac{\partial C}{\partial t} + \frac{\sigma_E^2 S^2}{2} \frac{\partial^2 C}{\partial S^2} + S(r_\$ - r_E) \frac{\partial C}{\partial S} - r_\$ C = 0,$$

where the $ and euro interest rate curves are $r_\$$, r_E, and the exchange rate follows the process

$$dS = S\mu_E \, dt + S\sigma_E \, dz(t).$$

Let

$$X = \frac{1}{S}, \quad P(X, t) = XC\left(\frac{1}{X}, t\right);$$

$$\Rightarrow \frac{\partial C(S, t)}{\partial S} = -X^2 \frac{\partial (P(X, t)/X)}{\partial X},$$

$$\frac{\partial^2 C(S, t)}{\partial S^2} = X^4 \frac{\partial^2 (P(X, t)/X)}{\partial X^2} + 2X^3 \frac{\partial (P(X, t)/X)}{\partial X};$$

$$\Rightarrow \frac{\partial C(S, t)}{\partial S} = -X \frac{\partial P(X, t)}{\partial X} + P(X, t), \quad \frac{\partial^2 C(S, t)}{\partial S^2} = X^3 \frac{\partial^2 P(X, t)}{\partial X^2};$$

$$\Rightarrow P(X,t): \quad \frac{\partial P(X,t)}{\partial t} + \frac{\sigma_E^2 X^2}{2} \frac{\partial^2 P(X,t)}{\partial X^2}$$

$$+ X(r_E - r_\$) \frac{\partial P(X,t)}{\partial X} - r_E P(X,t) = 0,$$

and noting that the process for X must be

$$dX = X(-\mu_E + \sigma_E)\,dt - X\sigma_E\,dz(t),$$

we see that the two volatilities are the same, because the minus sign may be absorbed into a new definition of the Wiener process. Only the drift is different.

Thus we have the same security, described by

$$P(X,T) = K \max\left(\frac{1}{K} - X, 0\right)$$

$$P(S,t): \quad \frac{\partial P}{\partial t} + \frac{\sigma_E^2 X^2}{2} \frac{\partial^2 P}{\partial S^2} + S(r_E - r_\$) \frac{\partial P}{\partial S} - r_E P = 0.$$

Obviously this security is just K puts on the $ written in euros, valued in euros, and struck at exchange rate $K' = 1/K$. Note that one quite remarkable feature: Although this security is *exactly* the same security looked at from a different point of view—that is, the trader who tracks P&L in dollars and the trader who tracks P&L in euros—the delta is different. Thus

$$\frac{\partial C(S,t)}{\partial S} = P(X,t) - X \frac{\partial P(X,t)}{\partial X}$$

$$\Rightarrow$$

$$\frac{\partial C(S,t)}{\partial S} \neq \frac{\partial P(X,t)}{\partial X}.$$

The two traders will hedge with slightly different amounts of short euros and long dollars.

This seemingly paradoxical first glance is resolved by noting that the two different traders track their risk in different currencies and Ito's lemma kicks off an extra drift term due to this. One values their portfolio in dollars and the other values their portfolio in euros; the same (stochastic) portfolio has a different average drift when measured in a different currency. Mathematically the chain rule ensures the two deltas are different.

A particularly illuminating picture is as follows: After trading a euro call option and then financing and hedging their respective positions, two traders have the following portfolios, each of total value *zero*, and written down with values in \$ followed by a valuation of Trader 2's portfolio in euros using the exchange rate $S = 1/X$;

	Call on euro	\$	euro
Trader 1 (\$)	C	$-(C - \Delta S)$	$-\Delta S$
	K puts on \$	\$	euro
Trader 2 (\$)	$-C$	$(C - \Delta S)$	ΔS
Trader 2 (euro)	$-XC = -KP'$	$KX\Delta_X = KX\frac{\partial P'}{\partial X}$	$K(P' - X\Delta_X)$

where C is a \$ denominated call on euros struck at K valued in \$ and P' is a euro denominated put struck at $1/K$ and valued in euros. The prime on the P is flagging the fact that P' is a function of X (and not S). This shows that, if trader 1 is U.S. based, the portfolio is long a call on euros which he bought with borrowed cash (\$) and the proceeds of a hedge that is short euros. If trader 2 tracks P&L in euros then his portfolio is short K puts on \$, and is hedged by buying \$. Thus, the two views are neatly reconciled by noting that both traders have the same trade on except that one is long and the other is short. One of the traders interprets the two additional portfolio positions as financing and hedging, while the other trader interprets them the other way round; as hedging and financing.*

6.5. BLACK-SCHOLES EQUATION IN MARTINGALE VARIABLES: HIDDEN SYMMETRY 2

Here we take into consideration the general variable change to switch to normal coordinates, meaning that the Black-Scholes equation becomes the heat equation. First, we change to maturity \$ for the option price expressed as

$$C(S, t) = K \exp\left(-\int_{s=t}^{s=T} r_\$(s)\, ds \right) C'(S, t).$$

We put in a factor of strike, only to strengthen the idea that C' is the number of zero-coupon bonds to which the option is equivalent. The function $C'(S, t)$ is a completely different function to $C(S, t)$ but I use the prime notation just to remind the reader it is a "transformation" of the option price, C. The Black-Scholes equation then becomes

$$\frac{\partial C'}{\partial t} + \frac{\sigma^2 S^2}{2}\frac{\partial^2 C'}{\partial S^2} + S(r_\$ - r_S)\frac{\partial C'}{\partial S} = 0.$$

This is the backward Kolmogorov equation for the stock price process, which is

$$dS = S(r_\$ - r_S)\, dt + S\sigma\, dz(t).$$

---

*Thanks to Jeff Miller for this illumination.

Then we introduce the martingale $m(S, t)$ with process

$$m(S, t) = \frac{S \exp\left(-\int_{s=t}^{s=T} r_S(s)\, ds\right)}{K \exp\left(-\int_{s=t}^{s=T} r_\$(s)\, ds\right)},$$

$$dm = m\sigma\, dz(t).$$

We find that $C''(m, t) = C'(S(m), t)$, where C' is a function of S and t while C'' is a (different) function of m and t while

$$\frac{\partial C''}{\partial t} + \frac{\sigma^2 m^2}{2} \frac{\partial^2 C''}{\partial m^2} = 0.$$

Then finally the normal variable, another martingale x, is introduced with process

$$x = \ln(m) - \frac{1}{2} \int_{s=t}^{s=T} \sigma(s)^2\, ds,$$

$$dx = \sigma\, dz(t).$$

This means that the function $C'''(x, t) = C''(m(x, t), t)$ satisfies

$$\frac{\partial C'''}{\partial t} + \frac{\sigma^2}{2} \frac{\partial^2 C'''}{\partial x^2} = 0.$$

Furthermore, the initial conditions may be rewritten and we find

$$C(S, T) = \max(S - K, 0),$$

$$C(S, t; K, T; \sigma, r_\$, r_S) = Ke^{-r_\$(T-t)} C'''(x, t),$$

$$x = \ln\left[\frac{Se^{-r_S(T-t)}}{Ke^{-r_\$(T-t)}}\right] - \frac{\sigma^2}{2}(T - t),$$

$$\Rightarrow C'''(x, T) = \max(\exp(x) - 1, 0),$$

$$C'''(x, t): \quad \frac{\partial C'''}{\partial(\sigma^2 t)} + \frac{1}{2} \frac{\partial^2 C'''}{\partial x^2} = 0,$$

where we have switched to showing (maturity) averaged interest rates and volatility; without the time integrals explicitly.

This is a remarkable result. It means:

- All rate dependence, stock and strike dependence is folded into the martingale function x—that is, a second hidden symmetry; the variable m, is actually the ratio of the stock forward and the strike zero price (cash forward),

- Interest rates and volatilities show up *only* as their averages over the life of the (European) option.
- The volatility shows up as a drift in the martingale x and as a scaling of the time, t.
- The price can be written as a function with *no* explicit rates dependence (an important result for later) as

$$C(S, t; K, T; \sigma, r_\$, r_S) = C'(Se^{-r_S(T-t)}, Ke^{-r_\$(T-t)}, \sigma^2(T-t)).$$

The generalization of this result is very useful. Given a pricing equation,

$$\Theta C(t) = 0,$$

where Θ represents the operator corresponding to the Black-Scholes operator

$$\Theta = \frac{\partial}{\partial t} + g_{11}(x_1)\frac{\partial^2}{\partial x_1^2} + g_{22}(x_2)\frac{\partial^2}{\partial x_2^2} + g_{12}(x_1, x_2, \ldots)\frac{\partial^2}{\partial x_1 \partial x_2} + \cdots$$

$$+ g_1(x_1, x_2, \ldots)\frac{\partial}{\partial x_1} + \cdots g_0(r_1, r_2, \ldots),$$

with two solutions, which may be a pricing equation over many state variables, $f_1(t)$ and $f_2(t)$, then any pricing solution may be written as

$$C(t) = f_2(t)F(m, t),$$

where $m = \frac{f_1(t)}{f_2(t)}$ and then $F(m, t)$ satisfies

$$\frac{\partial F}{\partial t} + g_m(m)\frac{\partial^2 F}{\partial m^2} = 0.$$

This can be seen because the two solutions we know, $f_1(t)$ and $f_2(t)$, correspond to $F = 1$ and $F = m$, and so there are no zero- or first-order terms. The second-order term must be of similar form to the original second-order term and can be determined by establishing the process for m.

6.6. BLACK-SCHOLES EQUATION WITH STOCK AS A "DERIVATIVE" OF OPTION PRICE: HIDDEN SYMMETRY 3

In this sixth section, a further hidden symmetry is due to the view that the underlying security may be the option and the derivative may be the stock.

To see this, we need to rewrite the Black-Scholes formula as a function S of C.

$$(S, t) \mapsto (S_2, t_2)$$

$$S = S(S_2, t_2); \quad t = t_2$$

$$\textit{where} \quad S_2 = C(S, t); \quad t_2 = t$$

$$\Rightarrow \frac{\partial}{\partial S_2} = \frac{\partial S}{\partial C} \frac{\partial}{\partial S}$$

$$\Rightarrow \frac{\partial}{\partial t_2} = \frac{\partial}{\partial t} + \frac{\partial S}{\partial t_2} \frac{\partial}{\partial S}.$$

Solving the last two equations to get the same functionality on either side,

$$\left(\frac{\partial S}{\partial C}\right)^{-1} \frac{\partial}{\partial S_2} = \frac{\partial}{\partial S},$$

$$\frac{\partial}{\partial t_2} - \frac{\partial S}{\partial t_2} \left(\frac{\partial S}{\partial C}\right)^{-1} \frac{\partial}{\partial S_2} = \frac{\partial}{\partial t},$$

we find that the Black-Scholes equation may be transformed to these new variables, under the transformation

$$\frac{\partial}{\partial S} \mapsto \left(\frac{\partial S}{\partial C}\right)^{-1} \frac{\partial}{\partial C},$$

$$\frac{\partial^2}{\partial S^2} \mapsto \left(\frac{\partial S}{\partial C}\right)^{-2} \frac{\partial^2}{\partial C^2} - \left(\frac{\partial S}{\partial C}\right)^{-3} \frac{\partial^2 S}{\partial C^2} \frac{\partial}{\partial C},$$

$$\frac{\partial}{\partial t} \mapsto \frac{\partial}{\partial t} - \frac{\partial S}{\partial t} \left(\frac{\partial S}{\partial C}\right)^{-1} \frac{\partial}{\partial C},$$

and, in operator notation, the Black-Scholes equation becomes:

$$\frac{\partial}{\partial t} + \frac{\sigma_S^2 S^2}{2} \frac{\partial^2}{\partial S^2} + S(r_\$ - r_S)\frac{\partial}{\partial S} - (r_\$ - r_C),$$

$$\mapsto$$

$$\frac{\partial}{\partial t} - \frac{\partial S}{\partial t} \left(\frac{\partial S}{\partial C}\right)^{-1} \frac{\partial}{\partial C} + \frac{\sigma_S^2 S^2}{2} \left(\frac{\partial S}{\partial C}\right)^{-2} \frac{\partial^2}{\partial C^2} - \frac{\sigma_S^2 S^2}{2} \left(\frac{\partial S}{\partial C}\right)^{-3} \frac{\partial^2 S}{\partial C^2} \frac{\partial}{\partial C}$$

$$+ S(r_\$ - r_S) \left(\frac{\partial S}{\partial C}\right)^{-1} \frac{\partial}{\partial C} - (r_\$ - r_C).$$

Letting this operate on what is now a simple coordinate C, we get only the coefficients of the $\frac{\partial}{\partial C}$ terms and the last multiplicative term:

$$\left[\frac{\partial}{\partial t} + \frac{\sigma_S^2 S^2}{2}\frac{\partial^2}{\partial S^2} + S(r_\$ - r_S)\frac{\partial}{\partial S} - (r_\$ - r_C)\right]C(S,t) = 0$$

$\mapsto$

$$-\frac{\partial S}{\partial t}\left(\frac{\partial S}{\partial C}\right)^{-1} - \frac{\sigma_S^2 S^2}{2}\left(\frac{\partial S}{\partial C}\right)^{-3}\frac{\partial^2 S}{\partial C^2} + S(r_\$ - r_S)\left(\frac{\partial S}{\partial C}\right)^{-1} - (r_\$ - r_C)S = 0.$$

Multiplying by $\frac{\partial S}{\partial C}$ and noting that $\frac{\partial S}{\partial C}\sigma_C C = \sigma_S S$, by comparing the processes for S and C, this formula defines the function $\sigma_C = \sigma_C(C,t)$. We finally obtain, after a change of sign and the last two terms reordered as

$$\frac{\partial S}{\partial t} + \frac{C^2\sigma_C^2}{2}\frac{\partial^2 S}{\partial C^2} + C(r_\$ - r_C)\frac{\partial S}{\partial C} - (r_\$ - r_S)S = 0.$$

We regain the Black-Scholes equation with a complete switch of S and C. This is hidden symmetry number 3; the fact that the formalism can be changed to switch the view of which security is the derivative and which is the underlying.

CHAPTER **7**

Interest Rate Hedging

B ond prices and interest rates are not fixed. We certainly might decide to use models with stochastic interest rates. Up to now, only stock prices (and currency exchange rates) have been modeled as stochastic processes. Before we do this, however, we may ask some simple questions about options such as:

- Given that we have a model of stock options and that the value of stock options is driven more by stock volatility than rates volatility, what might the interest rate hedging strategy look like?
- What is the recommended rates hedge implied by the model we have?

Strictly speaking, the model we have already developed for European options takes the volatility of the martingale—the ratio of stock forward to strike present value (PV)—as an input. This implies that stochastic rates, interest rates, and stock loan rates together with stochastic stock price have already been modeled. We merely need to reinterpret the volatility input. Now, the correct hedge is a combination of stock forward and a zero-coupon bond of maturity equal to option expiration. The following analysis shows how much zero-coupon bond to hedge with together with stock forward.

7.1. EULER'S RELATION

Given a call option trading at some price with a particular strike and stock price, consider the price of a related option with twice the strike price and the "underlying" is two shares of stock. This means that instead of one option we essentially have an inseparable package of two options. The price should usually be twice the single price. In the real world, it might be slightly less because we lose some optionality: When the prices get very high—the strike can be more than the GDP of any one country—the money supply in the market would be affected and the demand would drop as the ability of

62

any single market participant to afford the higher-priced option package is reduced. Ignoring this obviously academic caveat, we may write

$$C(\lambda S, \lambda K) = \lambda C(S, K).$$

Differentiating with respect to λ and then setting it to 1, we find Euler's relation

$$C(S, K) = K \frac{\partial C(S, K)}{\partial K} + S \frac{\partial C(S, K)}{\partial S}.$$

This result can easily be generalized to options with multiple products. It is an extremely useful result, particularly for discussions of optimal hedging strategies.

7.2. INTEREST RATE DEPENDENCE

From what we learned in the previous chapter (see section 6.5 Black-Scholes Equation in Martingale Variables: Hidden Symmetry 2), the functional form of rates dependence for European equity options is

$$C(S, K; \sigma, t, T; r_\$, r_S) = C'(Se^{-r_S(T-t)}, Ke^{-r_\$(T-t)}, \sigma^2(T-t)).$$

The rates here are average rates, meaning that the option is sensitive only to term interest rates that go to maturity of the option or, more generally, the cash flow date of the products. The sensitivity of the option to these rates may be seen, using the chain rule, to be

$$\frac{\partial C}{\partial r_\$} = \frac{\partial(Ke^{-r_\$(T-t)})}{\partial r_\$} \frac{\partial C}{\partial(Ke^{-r_\$(T-t)})} = \frac{\partial Z_K}{\partial r_\$} \Delta_{Z_K},$$

$$\frac{\partial C}{\partial r_S} = \frac{\partial(Se^{-r_S(T-t)})}{\partial r_S} \frac{\partial C}{\partial(Se^{-r_S(T-t)})} = \frac{\partial Z_S}{\partial r_\$} \Delta_{Z_S}.$$

Defining the deltas as the derivative's sensitivity to the respective zero-coupon bond (or "zero") prices, that is, PVs, for each cash flow,

$$\Delta_{Z_K} = \frac{\partial C}{\partial(Ke^{-r_\$(T-t)})},$$

$$\Delta_{Z_S} = \frac{\partial C}{\partial(Se^{-r_S(T-t)})}.$$

However, rewriting the Euler relation as

$$C(S, K) = Ke^{-r_\$(T-t)} \frac{\partial C(S, K)}{\partial (Ke^{-r_\$(T-t)})} + Se^{-r_\$(T-t)} \frac{\partial C(S, K)}{\partial (Se^{-r_\$(T-t)})}$$

$$= Z_K \Delta_{Z_K} + Z_S \Delta_{Z_S},$$

and comparing the two expressions, particularly the rates derivative of the second Euler equation, the result is that the first-order rates dependence is exactly replicated in a portfolio of zero-coupon bonds containing Z_K and Z_S (zero-coupon bonds in different currencies or products). The amounts of these bonds, Δ_{Z_K} and Δ_{Z_S}, may be called *zero deltas, cash flow deltas,* or *coupon deltas.*

The significant result is that for first-order rates dependence we can ignore the rates dependence of the coupon deltas. These terms add to zero as long as the Euler relation holds. The entire first-order rates dependence is exactly replicated by a portfolio containing these zeros in the amounts given by the corresponding coupon deltas.

Restricting to one product, U.S. dollars for example, and considering a convertible bond (CB), which for now we merely note is dependent on all the cash flows of the underlying bond and which is also a derivative of all these cash flows (i.e., coupons) of various dates and because the holder can convert to a given number of shares of a particular underlying stock at any time,

$$CB(S, c_1, \ldots, c_N; \sigma, t, T; r_\$, r_S)$$

$$= CB'(Se^{-r_\$(T-t)}; c_1 e^{-r_\$(t_1-t)}, \ldots, c_N e^{-r_\$(t_N-t)}; \sigma^2(T-t))$$

$$= CB'(Z_S; Z_1, \ldots, Z_N; \sigma^2(T-t))$$

where

$$Z_j = c_j e^{-r_\$(t_j-t)}.$$

Note that the first-order rates dependence is replicated by a portfolio of zeros, one for each coupon (i.e., cash flow), with each zero having a maturity that is the same as the coupon date—a PV that is the same as the PV of each coupon multiplied by an amount equal to the coupon delta. It is given by

$$PV_j = \Delta_{Z_j} Z_j = \Delta_{Z_j} e^{-r_\$(t_j-t)}$$

$$= \left. \frac{\partial CB(Z_S; Z_1, \ldots, \lambda Z_j, \ldots, Z_N; \sigma^2(T-t))}{\partial \lambda} \right|_{\lambda=1},$$

$$\Delta_{Z_j} = \frac{PV_j}{Z_j} = \frac{PV_j}{e^{-r_s(t_j - t)}}.$$

Thus we now have a portfolio of zero-coupon bonds that replicates the first-order rates dependence of the derivative exactly,

$$\Pi = \sum_j PV_j.$$

This result is extremely powerful because we can hedge rates derivatives with other rates derivatives very easily by simply using a zero-coupon bond analysis. The algorithm is as follows: Turn all derivatives into their equivalent zero-coupon bond portfolios and then the best possible hedge recommendation is to hedge with this portfolio of zeros. This hedges the derivative's interest rate risk against all curve shape changes to first order.

7.3. TERM-STRUCTURED RATES HEDGING: DURATION BUCKETING

The final problem to consider is that generally a large portfolio has many cash flow dates and we do not want to hedge against every interest rate term dependence—the reason being the cost.

A simple trick may be used. Say we want to hedge a portfolio of zeros of many different maturities out 10 or so years and the trader wants to hedge with only two bonds: of maturities 5 and 10 years. The two instruments' maturities and today's date determine three bucket dates—5 and 10 years because the hedging instruments have their largest cash flows on these dates—and a valuation date bucket, which means cash.

First, the hedging instruments are bonds and, obviously, bonds break into portfolios of zeros. Now we need to fit each zero into the three buckets, 0 (valuation date), 5, and 10 years. We note that for any particular zero less than 10 years, two zeros with maturities of the nearest two buckets' dates, with PVs proportional to the distance away from the original zero maturity, will have the same cash value and the same duration as the original zero.

$$Z = Z_1 + Z_2, \quad tZ = t_1 Z_1 + t_2 Z_2 : \quad t_2 > t > t_1$$

$$\Rightarrow Z_1 = Z \frac{t_2 - t}{t_2 - t_1}, \quad Z_2 = Z \frac{t - t_1}{t_2 - t_1}.$$

Figure 7.1 displays the PV of the cash flow versus time.

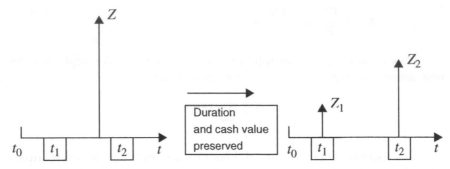

FIGURE 7.1   Duration bucketing of the present value of a cash flow into two nearby buckets.

Similarly we can change all the cash flows in the two hedging bonds to be duration bucketed into the three buckets, and thus represent the hedging instruments as collections of these three zeros.

We can similarly represent the portfolio as a collection of these three zeros. We can represent all derivatives as combinations of zeros and then we can bucket these zeros. If any cash flows in the portfolio are longer dated than the longest dated hedging instrument (the Nth hedging instrument for example), the best that can be achieved is duration matching to the last zero and adding in some valuation date cash to get the PV to match:

$$Z = Z_0 + Z_N, \quad (t - t_0)Z = (t_N - t_0)Z_2 : \quad t > t_N$$

$$\Rightarrow Z_N = Z\frac{t - t_0}{t_N - t_0}, \quad Z_0 = Z\left(1 - \frac{t - t_0}{t_N - t_0}\right).$$

The recommended hedge is the linear solution to the requirement that the total portfolio have no exposure to the two (forward) zeros at all. The remaining premium is all in the cash bucket and hence the portfolio has no rates exposure. This will uniquely determine the amounts of the two hedging instruments to short.

Consider that the portfolio does not strictly have no rates exposure. It just means that the hedged portfolio has the same exposure as cash, that is, the short-term (overnight) borrow rate of cash for financing the position. It means that this type of rates hedging is relevant for the trader or investor who is comparing his returns to riskless short-term borrow. Also it is not a hedge against arbitrary shape changes, just a *linear* hedge against independent parallel shifts of all the sections of the forward curve between each bucket date.

Because this duration bucketing preserves the cash PV of each zero and the first-order rates sensitivity (i.e., duration), it complements the above

Euler analysis well. That analysis also preserved the cash amount and the first-order rates sensitivity.

7.4. ALGORITHM FOR DECIDING WHICH HEDGING INSTRUMENTS TO USE

This leaves the last point: This algorithm says nothing about which hedging instruments to choose. To address this requires an analysis of the correlation matrix of all the various forward rates along the curve. Then, requiring that the variance of the total portfolio be less than some fixed value will uniquely fix the minimum subset of instruments with which to hedge from an arbitrary set of possible instruments.

CHAPTER 8

Interest Rate Derivatives: HJM Models

8.1. HULL-WHITE MODEL DERIVATION

8.1.1. Process and Pricing Equation

Stochastic interest rates add the extra dimension of trying to model a forward rate curve stochastically instead of just a stock price. The model is complicated by the fact that the first point of this curve is the risk-free (overnight, short, or spot) rate r that shows up in general risk-neutral pricing derivations.

We proceed with a simple arbitrage-free pricing model called the Hull-White model (following Cheyette 1992). Default-free zero bond prices $B_T(t)$ of maturity T at time t are written as

$$B_T(t) = e^{-\int_{s=t}^{s=T} r_s(t)\,ds},$$

which defines the instantaneous forward curve $r_T(t)$,

$$r_T(t) = -\frac{\partial}{\partial T}\ln B_T(t),$$

representing the cost of borrow from time T to $T + dT$, that may be *locked in* at time t. It is defined only for $T > t$. In other words, $r_T(t)$ is the forward curve observed at time t. The process for $r_T(t)$ is what is to be modeled and the initial value is the current forward curve (i.e., current zero-coupon bond prices),

$$r_T(0) = f(T).$$

Now, writing a generalized Brownian motion process for $B_T(t)$,

$$dB_T(t) = B_T(t)[\mu_T(t)\,dt + \sigma_T(t)\,dz(t)]$$

using a Wiener process $dz(t)$ together with arbitrary drift and volatility functions $\mu_T(t)$, $\sigma_T(t)$, the risk premium argument still applies (see section 4.1 Risk Premium Derivation) and determines that for market risk premium λ

$$\mu_T(t) = r_t(t) + \lambda\sigma_T(t).$$

At first sight this somewhat convoluted constraint might appear to be difficult to solve consistently.

First, as in Chapter 5 (see section 5.2 Derivation of Black-Scholes Equation, Risk-Neutral Pricing), the full derivation produces a differential equation that is not dependent on the value of the risk premium λ and so for simplicity we set it to zero.

We switch variables to find the process implied for $r_T(t)$, expressed as

$$r_T(t) = -\frac{\partial}{\partial T}\ln B_T(t)$$

$$\Rightarrow dr_T(t) = -\frac{\partial}{\partial T}\left[\frac{dB_T(t)}{B_T(t)} - \frac{1}{2}\sigma_T^2(t)\,dt\right],$$

where the second term is dictated by Ito's lemma. Thus

$$dr_T(t) = -\frac{\partial}{\partial T}\left[\mu_T(t)\,dt + \sigma_T(t)\,dz(t) + -\frac{1}{2}\sigma_T^2(t)\,dt\right]$$

$$= -\frac{\partial}{\partial T}\left[r_t(t)\,dt + \sigma_T(t)\,dz(t) + -\frac{1}{2}\sigma_T^2(t)\,dt\right]$$

$$= -\frac{\partial\sigma_T(t)}{\partial T}\,dz(t) + \frac{\partial\sigma_T(t)}{\partial T}\sigma_T(t)\,dt,$$

putting in $\lambda = 0$ and noting that the risk-less overnight rate is not dependent on the zero maturity. Clearly the process is far from simple. We can make some general observations about the (integrated) process for $r_T(t)$, that is,

$$r_T(t) = f(T) + \int_{s=0}^{s=t}\frac{\partial\sigma_T(s)}{\partial T}\sigma_T(s)\,ds - \int_{s=0}^{s=t}\frac{\partial\sigma_T(s)}{\partial T}\,dz(s).$$

This model is now a one-factor model, meaning that we introduced only one Wiener process, and so we can capture the model's characteristics by looking at the process for just the spot rate by setting $T - t$:

$$r_t(t) = f(t) + \int_{s=0}^{s=t}\frac{\partial\sigma_t(s)}{\partial t}\sigma_t(s)\,ds - \int_{s=0}^{s=t}\frac{\partial\sigma_t(s)}{\partial t}\,dz(s).$$

Defining $r = r_t(t)$ and then differentiating with respect to t, we find the process for the spot rate, expressed as

$$dr = \left[\frac{df(t)}{dt} + \sigma_t'(t)\sigma_t(t) + \int_{s=0}^{s=t} \sigma_t'(s)\sigma_t'(s) + \sigma_t''(s)\sigma_t(s)\,ds \right.$$
$$\left. - \int_{s=0}^{s=t} \sigma_t''(s)\,dz(s) \right] dt - \sigma_t'(t)\,dz(t)$$

where the primes refer to differentiation only with respect to the maturity argument,

$$\sigma_T'(t) = \frac{\partial \sigma_T(t)}{\partial T}$$

and we have grouped the terms for the spot process into drift and stochastic terms. A significant difficulty we have is valuing or simplifying the stochastic integral in the drift term.

To proceed we have to assume that the overnight rate has a volatility that may be written $\sigma(r, t)$ or, equivalently,

$$\sigma_t'(t) = \sigma(r, t).$$

We also need a much more restrictive assumption on the term structure of the volatility

$$\sigma_T'(t) = \sigma(r, t)\frac{F'(T)}{F'(t)}$$

with an arbitrary function $F(T)$ and $F'(T) = \frac{dF}{dT}$. Thus

$$\sigma_T(t) = \sigma(r, t)\frac{(F(T) - F(t))}{F'(t)}$$

and using this we can rewrite the forward curve value as

$$r_T(t) = f(T) + \int_{s=0}^{s=t} \frac{\partial \sigma_T(s)}{\partial T}\sigma_T(s)\,ds - \int_{s=0}^{s=t} \frac{\partial \sigma_T(s)}{\partial T}\,dz(s)$$

$$= f(T) + F'(T)\left(\int_{s=0}^{s=t} (F(T) - F(s))\left[\frac{\sigma(r,s)}{F'(s)}\right]^2 ds - \int_{s=0}^{s=t} \frac{\sigma(r,s)}{F'(s)}\,dz(s) \right).$$

Rearranging

$$-\int_{s=0}^{s=t} \frac{\sigma(r,s)}{F'(s)} \, dz(s) = \frac{r_T(t) - f(T)}{F'(T)} - \int_{s=0}^{s=t} (F(T) - F(s)) \left[\frac{\sigma(r,s)}{F'(s)}\right]^2 ds$$

and noting that the left-hand side does *not* depend on the maturity T (and therefore neither does the right-hand side in total), we can set $T = t$ to get

$$-\int_{s=0}^{s=t} \frac{\sigma(r,s)}{F'(s)} \, dz(s) = \frac{r - f(t)}{F'(t)} - \int_{s=0}^{s=t} (F(t) - F(s)) \left[\frac{\sigma(r,s)}{F'(s)}\right]^2 ds,$$

which in turn can be used to value the difficult term in the spot rate process:

$$-\int_{s=0}^{s=t} \sigma_t''(s) \, dz(s) = -\int_{s=0}^{s=t} \frac{\sigma(r,s) F''(t)}{F'(s)} \, dz(s)$$

$$= F''(t) \left(\frac{r}{F'(t)}\right) - F''(t) \int_{s=0}^{s=t} (F(t) - F(s)) \left[\frac{\sigma(r,s)}{F'(s)}\right]^2 ds.$$

Substituting this into the short rate process and canceling terms then implies that the short rate process is

$$d(r - f(t)) = \left[v^2(t) + \frac{d\ln[F'(t)]}{dt}(r - f(t))\right] dt - \sigma(r,t) \, dz(t)$$

where

$$v^2(t) = [F'(t)]^2 \int_{s=0}^{s=t} \left[\frac{\sigma(r,s)}{F'(s)}\right]^2 ds.$$

This is a mean reverting process for the short rate (as long as $\frac{d\ln[F'(t)]}{dt} < 0$, which implies that the volatility term structure is a decreasing function of time to maturity). This is a surprisingly simple result given that the only assumptions made were as follows:

- The model of bond price changes would be one factor.
- The short rate could be written as a generalized Brownian motion.
- The forward curve volatility could be written $\sigma_T'(t) = \sigma(r,t)\frac{F'(T)}{F'(t)}$.

The function $v^2(t)$ is essentially the integrated variance of paths. It is actually a term that has arisen in the same way as the extra term in

$$\langle \exp(x) \rangle = \exp \left(\langle x \rangle + \frac{1}{2}\mathrm{var}(x) \right)$$

mentioned a few times before in this text.

Finally, a further restricting choice of volatility term structure can simplify the analysis considerably. This choice is

$$F'(t) = e^{-kt},$$

$$\sigma(r,t) = \sigma_0$$

where

$$\tau = \frac{1}{k}$$

is a time scale over which the bond volatility goes to zero. Note how this implies that the short rate volatility is constant for all times and rates environments and that bond volatility is

$$\sigma_T(t) = \frac{\sigma_0}{k}(1 - e^{-k(T-t)}).$$

This then decays to zero as bonds mature. Qualitatively, at least, these assumptions are not entirely unreasonable. The biggest weakness is that normal rates imply that negative interest rates are possible even though the mean reverting process suppresses the likelihood of these as long as the short rate volatility is not too large compared to the initial forward curve. Note that the cumulative variance function is

$$v^2(t) = \sigma_0^2 \frac{(1 - e^{-2kt})}{2k}.$$

We have a very simple short rate process for this model, that is, the Hull-White model,

$$d(r - f(t)) = (v^2(t) - k(r - f(t)))\,dt - \sigma_0\,dz(t)$$

for which the pricing equation for all derivatives can be expressed as

$$C(r,t): \quad \frac{\partial C}{\partial t} + \frac{\sigma_0^2}{2}\frac{\partial^2 C}{\partial r^2} + (v^2(t) - k(r - f(t)) + f'(t))\frac{\partial C}{\partial r} - rC = 0.$$

Note $f'(t)$ is the first derivative of f with respect to t. This equation can be solved analytically in simple cases.

8.1.2. Analytic Zero-Coupon Bond Valuation

For zero-coupon bonds we write $x = r - f(t)$ and then the process and pricing differential equation are

$$dx = (v^2(t) - kx) \, dt - \sigma_0 \, dz(t),$$

$$\frac{\partial B}{\partial t} + \frac{\sigma_0^2}{2} \frac{\partial^2 B}{\partial x^2} + (v^2(t) - kx)\frac{\partial B}{\partial x} - (x + f(t))B = 0.$$

Then writing $B_T(x, t) = \exp\left(-a_T(t)(x) - b_T(t) - \int\limits_{S=t}^{S=T} f(s) \, ds\right)$, we find that the two functions must satisfy

$$-\frac{\partial a_T(t)}{\partial t} + ka_T(t) - 1 = 0, \quad a_T(T) = 0;$$

$$-\frac{\partial b_T(t)}{\partial t} + \frac{\sigma_0^2}{2}a_T(t)^2 - v^2(t)a_T(t) = 0, \quad b_T(T) = 0,$$

which implies

$$a_T(t) = \frac{1}{k}(1 - e^{-k(T-t)}),$$

$$b_T(t) = \frac{\sigma_0^2}{4k}a_T^2(t)(1 - e^{-2kt})$$

and thus

$$B_T(r, t) = \exp\left(-a_T(t)(r - f(t)) - b_T(t) - \int\limits_{S=t}^{S=T} f(s) \, ds\right).$$

This means that the $t = 0$ bond value is just $B_T(r = f(0), 0) = \exp\left(-\int_t^T f(s) \, ds\right)$. We have a one-factor model that is an almost mean-reverting to the forward-curve short-rate process model that has the following features:

- The corresponding bond price process is arbitrage free (and hence the extra drift term $v^2(t)$, which breaks the otherwise exact mean reversion).
- All bond prices, at time $t = 0$, are inputs. They are fixed using the current market prices and this implies an initial forward curve, $f(t)$.
- The process for $f(t)$ results in intermediate bond prices given by a variety of possible short rates r, but at maturity the bond price is exactly 1.

- The short rate has constant volatility σ_0, and the bonds have a long-dated volatility limit of σ_0/k, (equal to short rate volatility multiplied by a scale factor of time, $1/k$) that decays (over this scale factor of time) to zero as we go to maturity: $\sigma_T(t) = \frac{\sigma_0}{k}(1 - e^{-k(T-t)})$.
- These two parameters σ_0, k, are the model inputs.
- The forward curve value is

$$r_T(t) = f(T) + \frac{\sigma_0^2}{2k^2}\left[(e^{-kT} - 1)^2 - (e^{-k(T-t)} - 1)^2\right]$$

$$- \sigma_0 e^{-k(T-t)} \int\limits_{s=0}^{s=t} e^{-k(t-s)}\, dz(s).$$

- The average future values of the forward curve are near to the current forward curve—as we ride up the curve—but the second drift term takes the mean away from the exact current forward curve.
- The modes added to the forward curve as we go through time are exponentially suppressed as a function of distance along the curve,

$$dr_T(t) = \frac{\sigma_0^2}{k}(e^{-k(T-t)} - 1)e^{-k(T-t)}\, dt - \sigma_0 e^{-k(T-t)}\, dz(t).$$

That is, the spot rate has standard deviation σ_0 while the forward rates have a suppression of their standard deviation, $\sigma_0 e^{-k(T-t)}$.

8.1.3. Analytic Bond Call Option

An analytic expression for simple European call options, of maturity T_1 (where $0 < T_1 < T$), and strike K, on the same zero-coupon bond of maturity T we note that for

$$m = \frac{B_T(t)}{B_{T_1}(t)}.$$

Here m has a complicated process. But if we write the price of a call option as a product, expressed as

$$C(r, t) = B_{T_1}(r, t)F(m, t),$$

we find that $F(m, t)$ satisfies

$$\frac{\partial F}{\partial t} + \frac{m^2 \sigma_m^2}{2}\frac{\partial^2 F}{\partial m^2} = 0,$$

$$\sigma_m = |\sigma_T(t) - \sigma_{T_1}(t)| = \frac{\sigma_0}{k}(e^{-k(T_1-t)} - e^{-k(T-t)}).$$

This result is nearly obvious without doing all the algebra because the pricing equation written with r as the state variable has $B_T(t)$ and $B_{T_1}(t)$ as solutions, which correspond to $F(m, t) = m$ and $F(m, t) = 1$. Thus the equation of motion determining F can have only a second-order term, and this is determined by the process for m. The process for m is, not forgetting the Ito's lemma term,

$$dm = \frac{dB_T(t)}{B_{T_1}(t)} - \frac{B_T(t)}{B_{T_1}(t)^2} dB_{T_1}(t) + \frac{B_T(t)}{B_{T_1}(t)} \sigma_{T_1}^2(t) \, dt$$

$$= m\sigma_{T_1}^2(t) \, dt + m(\sigma_T(t) - \sigma_{T_1}(t)) \, dz(t).$$

The call option value is therefore writable using the Black-Scholes formula with the usual variables, $BS(S, K, T, \sigma, r_\$, r_{stock})$, as

$$C(r, t) = B_{T_1}(t) BS(m, K, T_1, \sigma_m, 0, 0),$$

$$m = \frac{B_T(t)}{B_{T_1}(t)},$$

$$\Sigma_m = \frac{1}{\sqrt{T_K}} \left(\int_{t=0}^{t=T_1} \sigma_m^2(t) dt \right)^{\frac{1}{2}} = \frac{\sigma_0}{k} \left(e^{-k(T - T_K)} - 1 \right) \sqrt{\frac{1 - e^{-2kT_K}}{2k}},$$

where the formulae for the bond prices are written explicitly in the previous section (see section 8.1.2 Analytic Zero-Coupon Bond Valuation).

For a fuller discussion of martingales, see section 6.5 Black-Scholes Equation in martingale Variables: Hidden Symmetry 2.

8.1.4. Calibration

Finally, note that calibrating such models contains an inherent ambiguity. It is due to the risk-neutral pricing paradigm. The pricing equation above corresponds to *not* one short-rate process but a whole (infinite) class of processes, namely, all processes of the form

$$dB_T(t) = B_T(t)[(r_t(t) + \lambda(t)\sigma_T(t)) \, dt + \sigma_T(t) \, dz(t)]$$

$$\Leftrightarrow \quad dr_T(t) = -\frac{\partial \sigma_T(t)}{\partial T} \, dz(t) + \frac{\partial \sigma_T(t)}{\partial T} (\sigma_T(t) - \lambda(t)) \, dt$$

for any value of the market risk premium λ. Even if this value itself has a stochastic process, the pricing equations and any formulae then derived are invariant. This remarkable feature, however, means that determining the process for the yield curve from analysis of actual data is very hard because separating the stochastic components of the curve from (possibly even stochastic) movements of the drift is very difficult.

8.2. ARBITRAGE-FREE PRICING FOR INTEREST RATE DERIVATIVES: HJM

The Hull-White model is a very simple example of a Heath-Jarrow-Morton (HJM) model after their seminal paper on these ideas (Heath, Jarrow, and Morton, 1992). These models address the issue of risk-neutral pricing and the resulting special drift term discussed above that shows up in the pricing equation for such models.

The basic question addressed by HJM is: What restrictions must be imposed on the dynamics of the term structure of interest rates that are consistent with arbitrage-free pricing and an arbitrary initial forward curve?

The process for the forward curve is assumed to be writable as the stochastic integral

$$r_T(t) = f(T) + \sum_{i=1}^{i=n} \int_{s=0}^{s=t} \Sigma_T^i(s)\,ds + \sum_{i=1}^{i=n} \int_{s=0}^{s=t} \sigma_T^i(s)\,dz_i(s),$$

where $r_T(t)$ is the instantaneous forward rate from time T to $T+dT$ observed at time t; $f(T)$ is the same forward curve observed at time $t = 0$; $dz_i(t)$, for $i = 1$ to n, are n independent Wiener processes, and $\sigma_T^i(t)$ and $\Sigma_T^i(t)$ are smooth functions. Ideally we would like to set the function $\Sigma_T^i(t)$ to zero to get a simple model of rates. But the central result of HJM is that this is not consistent with arbitrage-free pricing.

Consider first the following definitions. The spot rate process $r(t)$, given by $r_t(t)$, is implicitly defined by the stochastic integral

$$r(t) = r_t(t) = f(t) + \sum_{i=1}^{i=n} \int_{s=0}^{s=t} \Sigma_t^i(s)\,ds + \sum_{i=1}^{i=n} \int_{s=0}^{s=t} \sigma_t^i(s)\,dz_i(s);$$

the value of a money market account of value 1 at time $t = 0$ is

$$B(t) = \exp\left(\int_{s=0}^{s=t} r(s)\,ds\right),$$

and the time t value of a zero-coupon bond of maturity T is

$$Z_T(t) = \exp\left(-\int_{s=t}^{s=T} r_s(t)\,ds\right).$$

Note that any initial zero-coupon bond prices—the forward curve observed today—are completely freely specifiable, and may be expressed as

$$Z_T(0) = \exp\left(-\int_{s=0}^{s=T} f(s)\,ds\right).$$

This curve is the input to HJM models.

The processes for the forward curve and spot rate are derivable as

$$dr_T(t) = \sum_{i=1}^{i=n} \Sigma_T^i(t)\,dt + \sum_{i=1}^{i=n} \sigma_T^i(t)\,dz_i(t)$$

$$dr = \left[\frac{df(t)}{dt} + \sum_{i=1}^{i=n} \Sigma_t^i(t) + \sum_{i=1}^{i=n}\int_{s=0}^{s=t} \frac{\partial \Sigma_t^i(s)}{\partial t}\,ds + \sum_{i=1}^{i=n}\int_{s=0}^{s=t} \frac{\partial \sigma_t^i(s)}{\partial t}\,dz_i(s)\right]dt$$

$$+ \sum_{i=1}^{i=n} \sigma_t^i(t)\,dz_i(t).$$

The zero-coupon bond price process is then derivable as

$$dZ_T(t) = Z_T(t)\left[\{r(t) + \Theta\}\,dt + \sum_{i=1}^{i=n} a_T^i(t)\,dz_i(t)\right]$$

where

$$\Theta_T(t) = \sum_{i-1}^{i=n}\left\{\frac{1}{2}\left[\int_{s=t}^{s=T} \sigma_s^i(t)\,ds\right]^2 - \int_{s=t}^{s=T} \Sigma_s^i(t)\,ds\right\},$$

$$a_T^i(t) = -\int_{s=t}^{s=T} \sigma_s^i(t)\,ds.$$

Now arbitrage freedom (see section 4.1 Risk Premium Derivation), applied to the traded zero-coupon bonds, dictates that this term must be equal to a sum of constants times the various volatilities of each bond that trades

$$\Theta_T(t) = \sum_{i=1}^{i=n} a_T^i(t)\lambda_i.$$

Furthermore, for pricing (see section 5.2 Derivation of Black-Scholes Equation: Risk-Neutral Pricing) we may set the values of these constants to zero, that is,

$$\sum_{i=1}^{i=n} \left\{ \frac{1}{2} \left[\int_{s=t}^{s=T} \sigma_s^i(t)\, ds \right]^2 - \int_{s=t}^{s=T} \Sigma_s^i(t)\, ds \right\} = 0,$$

and so (differentiating with respect to T) we find that one possible solution is

$$\Sigma_T^i(t) = \left[\int_{s=t}^{s=T} \sigma_s^i(t)\, ds \right] \sigma_T^i(t).$$

Noting that $\Sigma_t^i(t) = 0$, the spot rate process simplifies by losing one term:

$$dr = \left[\frac{df(t)}{dt} + \sum_{i=1}^{i=n} \int_{s=0}^{s=t} \frac{\partial \Sigma_t^i(s)}{\partial t}\, ds + \sum_{i=1}^{i=n} \int_{s=0}^{s=t} \frac{\partial \sigma_t^i(s)}{\partial t}\, dz_i(s) \right] dt + \sum_{i=1}^{i=n} \sigma_t^i(t)\, dz_i(t).$$

For one-factor models of the forward curve, we might find the spot rate process and use the spot rate as a state variable. This is what happened in section 8.1 Hull–White Model Derivation. For two-factor models, we might find the process for the forward curve (short-end) slope, expressed as

$$g(t) = \left. \frac{\partial r_T(t)}{\partial T} \right|_{T=t}$$

and use $r(t)$ and $g(t)$ as state variables and derive a pricing equation in the two variables r and g. Models with higher numbers of factors or stochastic drivers may be constructed similarly. Often Monte Carlo methods are used instead of pricing differential equations because the numerical solutions to higher dimensional partial differential equations can be prohibitively slow.

CHAPTER 9

Differential Equations, Boundary Conditions, and Solutions

9.1. BOUNDARY CONDITIONS AND UNIQUE SOLUTIONS TO DIFFERENTIAL EQUATIONS

We should briefly make an important aside. The main question to consider is this: How many boundary conditions are required to uniquely specify a single solution to a given specific differential equation?

Second-order differential equations, meaning equations that contain derivatives no higher than second derivatives, fit into three types and they are, with examples:

- *Elliptic differential equations* such as the Laplace equation:

$$\frac{\partial^2 f}{\partial x^2} + \frac{\partial^2 f}{\partial y^2} + \cdots = 0.$$

- *Hyperbolic differential equations*, the wave equation:

$$-\frac{\partial^2 f}{\partial t^2} + \frac{\partial^2 f}{\partial x^2} + \frac{\partial^2 f}{\partial y^2} + \cdots = 0.$$

- Parabolic differential equations, of which the heat equation is best known:

$$-\frac{\partial f}{\partial t} + \frac{\partial^2 f}{\partial x^2} + \frac{\partial^2 f}{\partial y^2} + \cdots = 0.$$

In each case the ellipsis dots signify other functions and terms not containing derivatives higher than second derivatives.

To classify differential equations in *two* variables, write them as

$$L(f) = a(x,t)\frac{\partial^2 f}{\partial t^2} + 2b(x,t)\frac{\partial^2 f}{\partial x \partial t} + c(x,t)\frac{\partial^2 f}{\partial x^2} + g\left(\frac{\partial f}{\partial x}, \frac{\partial f}{\partial t}, x, t\right) = 0.$$

Then change variables to find the *normal form*, i.e., switch to a new coordinate system that makes the coefficients of the second-order derivatives into simple numbers, and three types emerge as follows:

1. Elliptic if $a(x,t)c(x,t) - b(x,t)^2 > 0$ (e.g., Laplace equation)
2. Hyperbolic if $a(x,t)c(x,t) - b(x,t)^2 < 0$ (e.g., wave equation)
3. Parabolic if $a(x,t)c(x,t) - b(x,t)^2 = 0$ (e.g., heat equation and Black-Scholes)

To get a unique *finite* valued function f as a solution to

$$a(x,t)\frac{\partial^2 f}{\partial t^2} + 2b(x,t)\frac{\partial^2 f}{\partial x \partial t} + c(x,t)\frac{\partial^2 f}{\partial x^2} + g\left(\frac{\partial f}{\partial x}, \frac{\partial f}{\partial t}, x, t\right) = 0$$

over a region S, of the (x,t) plane, bounded by ∂S say, requires the following in each case:

- For elliptic, an arbitrary (boundary value) function specifying the value of the function f all the way around the closed boundary of the region S.
- For hyperbolic across *one* space-like slice (e.g., $x = 0$), the function f, and its derivative, $\frac{\partial f}{\partial t}$, need to be specified, and the same along the time-like boundaries of the region. This determines the function uniquely in a wide region forward and backward in time. (This region obviously needs to include the required region of uniqueness, S for the function, f.)
- For parabolic, an arbitrary (boundary value) function along any three sides of the region (e.g., $f(x,T)$) over x at time T; $f(+R,t)$ over t along the upper edge of the region; and $f(-R,t)$ over t along the lower edge of the region solving the differential equation subject to these boundary conditions uniquely specifies the function within the region *and* along the remaining side $t = 0$.

Furthermore, we can interpret these conditions.

1. Elliptic equations are the spatial part of a 2-D (or more) heat equation with no temperature transfer. Thinking physically in 2-D, if we specify the temperature around the edge of a piece of metal plate, then after a long time a steady-state temperature distribution over the whole plate is reached. This temperature distribution is unique and corresponds to the unique solution of an elliptic differential equation.

2. Hyperbolic equations describe wave propagation, and we need to specify the amplitude and the gradient of the wave at time $t = constant$, and then the wave propagates out at a speed that is implicit in the equation.
3. Parabolic equations are the most general heat equation with diffusion. If we specify the initial heat distribution at time $t = 0$ over some region, together with the temperature (or heat transfer, i.e., temperature gradient) at the edges of the region at all future times, then the temperature distribution over the region at all times is unique.

These are special cases of the so-called *Cauchy problem*, finding solutions to differential equations given boundary conditions. There are more solutions if we allow the functions to diverge at points within the region R. These functions, however, generally do not have much relevance to finance. Here we focus only on parabolic equations (the heat equation) in finance due to the stochastic nature of the processes used to model security prices.

Much effort, of course, has been expended on solving differential equations, particularly of second order, in the last three centuries and many textbooks are available with varying depths of detail. A good general text is, *Methods of Mathematical Physics* by Richard Courant and David Hilbert.

9.2. SOLVING THE BLACK-SCHOLES OR HEAT EQUATION ANALYTICALLY

9.2.1. Green's Functions

The analytical solution (i.e., integration) of differential equations requires the development of tools to help us. Two of the most useful follow. First, let us discuss the Black-Scholes (heat) equation in normal variables in one spatial dimension, expressed as

$$\frac{\partial f}{\partial t} + \frac{\sigma^2}{2}\frac{\partial^2 f}{\partial x^2} = 0.$$

If we are given the boundary conditions, beginning with the terminal price distribution or initial temperature,

$$f(x, T) = f_T(x),$$

we then note two properties of the following function:

$$P(x, t; x_T, T) = \frac{1}{\sqrt{2\pi\sigma^2(T-t)}}\exp\left[-\frac{(x-x_T)^2}{2\sigma^2(T-t)}\right].$$

It solves the equation (in x and t); and at time $t = T$ it is a Dirac delta function. These two properties mean that it may be used as a Green's function or propagator. The result of the (convolution) integral is

$$f(x, t) = \int_{x=-\infty}^{x=\infty} f_T(x_T) P(x, t; x_T, T) \, dx_T$$

which solves the heat equation everywhere because P does and at time $t = T$

$$f(x, T) = f_T(x)$$

because P is a Dirac delta at this time. Note that there are boundary conditions at $x \mapsto \pm\infty$ for all intermediate times. They are implicit in the P solution. In fact, there are other solutions besides P that are Dirac deltas at $t = T$, but not zero at $x \mapsto \pm\infty$.

We already used this method to obtain the Black-Scholes formula in Chapter 4 (see section 4.2 Analytic Formula for the Expected Payoff of a European Option). Formally, a Green's function for a particular differential operator, say L, satisfies

$$L(x, y, \ldots, t) G(x - x_0, y - y_0, \ldots, t - t_0;) = \delta(x - x_0)\delta(y - y_0)\ldots\delta(t - t_0),$$

which implies for this equation that

$$\frac{\partial G(x - x_0, t - t_0)}{\partial t} + \frac{\sigma^2}{2} \frac{\partial^2 G(x - x_0, t - t_0)}{\partial x^2} = \delta(t - t_0)\delta(x - x_0).$$

Solving this (by 2-D Fourier transform, see Appendix B), it can be shown that

$$G(x, t; x_T, T) = H(T - t) \frac{1}{\sqrt{2\pi\sigma^2(T - t)}} \exp\left[-\frac{(x - x_T)^2}{2\sigma^2(T - t)}\right].$$

Where $H(x)$ is the unit step function

$$H(x) = \begin{cases} 1 & x > 0 \\ 0 & x < 0 \end{cases},$$

and thus the Green's function is defined only for $t < T$.

9.2.2. Separation of Variables

The second method is to look for a solution to

$$\frac{\partial f}{\partial t} + \frac{\sigma^2}{2}\frac{\partial^2 f}{\partial x^2} = 0$$

of the form

$$f(x,t) = \Psi(x)\Theta(t),$$

which implies

$$\frac{1}{\Theta(t)}\frac{\partial\Theta(t)}{\partial t} = -\frac{1}{\Psi(x)}\frac{\partial^2\Psi(x)}{\partial x^2}.$$

This can be solved only if both terms equal a constant, say k^2. That is, we have separated the variables,

$$\frac{\partial\Theta_k(t)}{\partial t} = \left(\frac{\sigma^2}{2}k^2\right)\Theta_k(t)$$

$$\frac{\partial^2\Psi_k(x)}{\partial x^2} = -k^2\Psi_k(x).$$

The solution

$$f_k(x,t) = \Theta_k(t)\Psi_k(x)$$

$$= \exp\left[-\frac{\sigma^2 k^2(T-t)}{2}\right]\exp[ikx]$$

solves the heat equation for all $k: -\infty < k < \infty$ and so we can find a solution, subject to the boundary condition if we can solve

$$f(x,T) = f_T(x) = \frac{1}{\sqrt{2\pi}}\int_{k=-\infty}^{k=\infty} f_T(k)\exp(ikx)\,dk$$

because the full solution is then

$$f(x,t) = \frac{1}{\sqrt{2\pi}}\int_{k=-\infty}^{k=\infty} f_T(k)\exp\left(ikx - \frac{\sigma^2 k^2(T-t)}{2}\right)dk.$$

The solution is thus obtained by (1-D) Fourier transformation of the boundary function and then inverse (1-D) Fourier transformation of the product of this and the function

$$\Theta_k(t) = \exp\left[-\frac{\sigma^2 k^2(T-t)}{2}\right].$$

Again, there are boundary conditions implicit at large $\pm x$. Note that the Fourier transform of the call option payout is

$$x = \ln(S/K) + \left(r_\$ - r_{st} - \frac{\sigma^2}{2}\right)(T-t),$$

$$f_T(k) = \frac{1}{\sqrt{2\pi}} \int_{x=-\infty}^{x=\infty} f_T(x) \exp(-ikx)\, dx$$

$$= \frac{K}{\sqrt{2\pi}} \int_{x=-\infty}^{x=\infty} \max(e^x - 1, 0) \exp(-ikx)\, dx$$

$$= \frac{K}{\sqrt{2\pi}} \int_{x=0}^{x=\infty} (e^x - 1) \exp(-ikx)\, dx$$

$$= \frac{K}{\sqrt{2\pi}} \left(\frac{-1}{1-ik} + \frac{i}{k}\right).$$

Thus the Fourier transform of the Black-Scholes call option formula immediately follows

$$C(k, T - t; K) = \frac{K}{\sqrt{2\pi}} \left(\frac{-1}{1-ik} + \frac{i}{k}\right) \exp\left(-\frac{\sigma^2 k^2 (T-t)}{2}\right).$$

Clearly, this is probably not the easiest method to use for the case of the Black-Scholes equation! Nevertheless, the inverse Fourier transform does give the Black-Scholes call formula as

$$C(x(S, K, T - t, \sigma, r_\$, r_S), t) = \frac{1}{\sqrt{2\pi}} \int_{k=-\infty}^{k=\infty} C(k, T - t; K) \exp(ikx)\, dk.$$

9.3. SOLVING THE BLACK-SCHOLES EQUATION NUMERICALLY

9.3.1. Finite Difference Methods: Explicit/Implicit Methods, Variable Choice

Let's rewrite the Black-Scholes equation

$$\frac{\partial C}{\partial t} + \frac{\sigma^2}{2} \frac{\partial^2 C}{\partial x^2} + \left(r_\$ - r_S - \frac{\sigma^2}{2}\right) \frac{\partial C}{\partial x} - r_\$ C = 0,$$

in discrete form on a (time and stock-price space) grid as

$$\frac{(C_{i,j} - C_{i-1,j})}{(t_i - t_{i-1})} + \frac{\sigma_{i,j}^2}{2} \frac{(C_{i,j+1} - 2C_{i,j} + C_{i,j-1})}{(x_{j+1} - x_j)(x_j - x_{j-1})}$$

$$+ \left(r_{\$;i} - r_{S;i} - \frac{\sigma_{i,j}^2}{2}\right) \frac{(C_{i,j+1} - C_{i,j})}{(x_{j+1} - x_j)(x_j - x_{j-1})} - r_\$ C_{i,j} = 0.$$

This may be written

$$C_{i-1,j} = C_{i,j} + (t_i - t_{i-1}) \left[\frac{\sigma_{i,j}^2}{2} \frac{(C_{i,j+1} - 2C_{i,j} + C_{i,j-1})}{(x_{j+1} - x_j)(x_j - x_{j-1})} \right.$$

$$\left. + \left(r_{\$;i} - r_{S;i} - \frac{\sigma_{i,j}^2}{2}\right) \frac{(C_{i,j+1} - C_{i,j})}{(S_{j+1} - S_j)(S_j - S_{j-1})} - r_\$ C_{i,j} \right]$$

$$= p_{j+1} C_{i,j+1} + p_j C_{i,j} + p_{j-1} C_{i,j-1}.$$

Thus we find an algorithm to solve this parabolic second-order differential equation in two variables by moving one step backward in the time direction iteratively. The formula is a linear combination of the three nearest points at the current time-step. Generally, the algorithm consists of specifying boundary conditions on the three sides of the grid and then using this iterative method to find a numerical approximation for the solution on the third edge of the grid. It is obviously very simple and straightforward.

This is not the only way to discretize the differential equation. We could symmetrize the drift term (or eliminate it by choosing drifting variables) and, more importantly, we could value the spatial derivatives at time $i - 1$ instead of at time i as we did here. We would get a formula in this case that has the form

$$p_{j+1} C_{i-1,j+1} + p_j C_{i-1,j} + p_{j-1} C_{i-1,j-1} = C_{i,j}.$$

Now the previous simplicity is replaced by the requirement to solve a linear set of equations at each time-step. Methods like these are called *implicit* methods and the former are called *explicit* methods. It turns out that implicit methods have much better convergence properties and require fewer time-steps and therefore are of higher speed for the same accuracy level. Explicit methods are much easier to implement, maintain, and develop. For complicated financial derivatives, such as when the boundary conditions change every day, the extra speed usually obtained in implicit methods does not help because the security's specs require testing at every time-step. (See, for example, section 9.3.5 American Exercise in this chapter.)

The first question to consider is this: Does this method converge to the full solution of the differential equation? Second, are we constrained in how we choose the weights—given that there are many ways to discretize the equation?

Applying the central limit theorem, we can test for correct convergence. We focus on the fully explicit method only. Any solution may be obtained by summing the components of the Fourier transform of the payout multiplied by a propagator function for each component (as discussed in section 9.2.2 Separation of Variables):

$$\text{Payout} = f(x, T) = \sum_k a_k \Psi_k(x),$$

$$\text{Full solution} = f(x, t) = \sum_k a_k f_k(x, t);$$

$$f_k(x, t) = \Theta_k(t)\Psi_k(x);$$

$$\Theta_k(t) = \exp[-r_\$(T - t)] \exp\left[-\frac{\sigma^2 k^2 (T - t)}{2} + ik(r_{short} - \sigma^2/2)(T - t)\right];$$

$$\Psi_k(x) = \exp[ikx].$$

Now, put each component on the grid above and write the propagation formula in the following way for simplicity as

$$C_{i-1,j} = \exp(-r_\$\Delta t)(p_{j+1}C_{i,j+1} + p_j C_{i,j} + p_{j-1}C_{i,j-1}).$$

After one grid time-step, the Fourier components of the payout function are transformed as follows:

$$f_k(x_j, T) = \exp(ikx_j)$$

$$f_k(x_j, T - \Delta t) = \exp(ikx_j)\exp(-r_\$\Delta t)(p_{j+1}e^{+ik\Delta x} + p_j + p_{j-1}e^{-ik\Delta x}).$$

Utilizing the central limit theorem (see Appendix A: Central Limit Theorem-Plausibility Argument), and assuming that the sum of the weights is *1*, we find the limit of $N \to \infty$ grid time-steps,

$$f_k(x_j, T) = \exp(ikx_j),$$

$$f(x_j, T - N\Delta t) \xrightarrow{N \to \infty} \exp(-r_s N\Delta t) \int_{y=-\infty}^{y=+\infty} \frac{1}{\sqrt{2\pi \Sigma^2 N}}$$

$$\exp\left(-\frac{(y + NM)^2}{2\Sigma^2 N}\right) \exp(ik(x_j - y)) \, dy$$

$$= \exp(ikx_j) \exp(-r_s N\Delta t) \exp[ikMN] \exp\left[-\frac{k^2 \Sigma^2 N}{2}\right],$$

$$M = (p_{j+1} - p_{j-1})\Delta x,$$

$$\Sigma^2 = (p_{j+1} + p_{j-1})\Delta x^2 - M^2,$$

with the caveat that the grid must have even spacing in x-space, otherwise the central limit theorem would give a different limit. (The central limit theorem picks out the variables where the state density is constant.) For example, here we used a grid evenly spaced in $\ln(S)$ space. This resulted in a normal distribution of stock returns and, correspondingly, a lognormal terminal stock price distribution. But if we used a constant grid-spacing in stock price space, we would get normal stock prices in the limit of step size goes to zero. This is obviously a very important issue because it might look like the differential equation is being solved by discretizing but the limit solves a different equation.

Furthermore, comparing the full solution and this central limit result after many time-steps, correct convergence requires that as the step sizes go to zero

$$p_{j+1} + p_j + p_{j-1} = 1,$$

$$ik(p_{j+1} - p_{j-1})N\Delta x = +ik(r_{short} - \sigma^2/2)N\Delta t,$$

$$-\frac{k^2 N}{2}\{(p_{j+1} + p_{j-1})\Delta x^2 - [(p_{j+1} - p_{j-1})\Delta x]^2\} = -\frac{\sigma^2 k^2 N\Delta t}{2},$$

or

$$p_{j+1} + p_j + p_{j-1} = 1,$$

$$(p_{j+1} - p_{j-1})\Delta x = (r_{short} - \sigma^2/2)\Delta t,$$

$$(p_{j+1} + p_{j-1})\Delta x^2 = \sigma^2 \Delta t + O(\Delta t^2).$$

where $O(\Delta t^2)$ is a term that goes to zero as Δt^2. i.e. $\lim \frac{O(\Delta t^m)}{\Delta t^{m-1}} = 0$ for any m.

These constraints are the first three moments of the distribution. So, in summary, if we apply the first three moments as constraints on the weights and make sure the grid is evenly spaced in normal variable space, we get the solution to the differential equation in the small grid-step-size limit for each Fourier mode of the terminal distribution and therefore for any terminal distribution.

Now, we will actually choose the three weights, p_{j+1}, p_j, p_{j-1} as follows:

$$C_{i-1,j} = \exp(-r_\$ \Delta t)(p_{j+1} C_{i,j+1} + p_j C_{i,j} + p_{j-1} C_{i,j-1}),$$

$$C(S, t) = K \exp(-r_\$(T - t))$$

$$\Rightarrow 1 = p_{j+1} + p_j + p_{j-1},$$

$$C(S, t) = S \exp(-r_S(T - t)) = S \exp(-r_S(T - t)) \exp(-r_{short}(T - t))$$

$$\Rightarrow e^{-r_{short} \Delta t} = p_{j+1} e^{\Delta x} + p_j + p_{j-1} e^{-\Delta x},$$

$$C(S, t) = S^2 \exp(-r_S(T - t)) \exp(\sigma^2(T - t) + 2r_{short}(T - t))$$

$$\Rightarrow e^{\sigma^2 \Delta t + 2r_{short} \Delta t} = p_{j+1} e^{2\Delta x} + p_j + p_{j-1} e^{-2\Delta x}$$

where $r_{short} = r_\$ - r_S$ and $S = e^x$.

Choosing the weights this way results in special properties. These are the first three moments of the stock price change distribution and they agree with the constraints above to first order. Note that this choice means that cash and stock forwards are valued *exactly* no matter how few time-steps are taken. The last constraint calibrates the stock price second moment, i.e., volatility.

We choose to do this instead of using the moments in stock-price-return space because these constraints are much more useful for finite step sizes: most derivatives are "simple" functions of stock, like stock options, not simple functions of ln(S). Further, the time slice boundary conditions (the three sides of a region require boundary conditions for a parabolic second-order differential equation) are consistent with this because, for the majority of derivatives, the boundary conditions are a linear sum of stock forwards and zeros. For example, the solution for stock call options deep in the money is

$$C(S, t) \xrightarrow[S >> K]{} S \exp(-r_S(T - t)) - K \exp(-r_\$(T - t)).$$

The solution for the weights is

$$p_{j-1} = \left(-\alpha e^{\Delta x} + e^{\frac{\Delta x}{2}} \beta\right) / \det,$$

$$p_{j+1} = \left(-\alpha e^{-\Delta x} + e^{\frac{-\Delta x}{2}} \beta\right) / \det,$$

$$p_j = 1 - p_{j+1} - p_{j-1},$$

where

$$\alpha = 2\sinh(\Delta x)(e^{r_{short}\Delta t} - 1),$$

$$\beta = 2\sinh\left(\frac{\Delta x}{2}\right)(e^{(\sigma^2 + 2r_{short})\Delta t} - 1),$$

$$\det = 2\sinh(2\Delta x) - 4\sinh(\Delta x).$$

Note that the process drift cannot be too great, to ensure that the average stock price drift for one time step lies within the three nearest space steps. This gives a lower bound on the space step sizes, expressed as

$$\Delta x > (r - r_S)\Delta t.$$

Finally, note the much simpler formulation in drifting variables and maturity dollars:

$$\frac{\partial C}{\partial t} + \frac{\sigma^2}{2}\frac{\partial^2 C}{\partial x^2} = 0.$$

The three simplest moment constraints are as follows:

$C_{i-1,j} = (p_{j+1}C_{i,j+1} + p_j C_{i,j} + p_{j-1}C_{i,j-1})$:

moment 0 : $\check{C}(x,t) = 1 \Rightarrow 1 = p_{j+1} + p_j + p_{j-1}$,

moment 1 : $C(x,t) = x \Rightarrow x = p_{j+1}(x + \Delta x) + p_j x - p_{j-1}(x - \Delta x)$,

$\Rightarrow p_{j+1} = p_{j-1}$

moment 2 : $C(x,t) = x^2 + \sigma^2(T - t) \Rightarrow \sigma^2\Delta t = p_{j+1}\Delta x^2 + p_{j-1}\Delta x^2$,

$$\Rightarrow p_{j+1} + p_{j-1} = \frac{\sigma^2\Delta t}{\Delta x^2},$$

$$p_{j+1} = p_{j-1} = \frac{\sigma^2\Delta t}{2\Delta x^2},$$

$$p_j = 1 - \frac{\sigma^2\Delta t}{\Delta x^2}.$$

For stability, the weights all need to be positive, thus

$$\Delta x^2 > \sigma^2\Delta t.$$

This means that for a given time-step we want the space steps to be separated by at least one standard deviation (less than this and the diffusion will not occur smoothly). There is no upper limit on time-steps due to the drift of the process as before.

9.3.2. Gaussian Kurtosis (and Skew = 0), Faster Convergence

We still have the freedom to choose the size of the stock price space grid-steps and the time grid-steps. We may ask: They both need to go to zero to get the correct limit, but can we choose a relation between them to get faster convergence? Von-Neumann stability mode analysis provides an answer.

Consider the simpler driftless equation of motion first.

$$\frac{\partial C}{\partial t} + \frac{\sigma^2}{2}\frac{\partial^2 C}{\partial x^2} = 0,$$

$$C_{i-1,j} = (p_{j+1}C_{i-1,j+1} + p_j C_{i-1,j} + p_{j-1}C_{i-1,j-1}),$$

$$p_{j+1} = p_{j-1} = \frac{\sigma^2 \Delta t}{2\Delta x^2}, \quad p_j = 1 - \frac{\sigma^2 \Delta t}{\Delta x^2}.$$

Look for a mode on the grid of the form, that is,

$$C_{i,j} = \zeta^{\frac{T}{\Delta t}} e^{ikx},$$

$$p = \frac{\sigma^2 \Delta t}{2\Delta x^2},$$

$$\zeta = 1 - 2p + 2p\cos(k\Delta x) = 1 - 4p\sin^2\left(\frac{k\Delta x}{2}\right).$$

We find that

$$\lim_{\substack{\Delta t \to 0 \\ p=const.}} \zeta^{\frac{T}{\Delta t}} = \lim_{\substack{\Delta t \to 0 \\ p=const.}} \left[1 - 4p\sin^2\left(\frac{k\Delta x}{2}\right)\right]^{\frac{T}{\Delta t}}$$

$$= \lim_{\substack{\Delta t \to 0 \\ p=const.}} \left[1 - \frac{\sigma^2 k^2 \Delta t}{2}\right]^{\frac{T}{\Delta t}} = \exp\left(-\frac{k^2\sigma^2 T}{2}\right),$$

which *is* the correct limit, but we can ask whether we can force convergence to be faster with a particular choice of p. Expand the mode to find

$$\zeta = 1 - 4p\sin^2\left(\frac{k\Delta x}{2}\right)$$

$$= 1 - pk^2\Delta x^2 + p\frac{(k\Delta x)^4}{12} + \cdots$$

and compare to the expansion of the full solution,

$$\exp\left(-\frac{k^2\sigma^2\Delta t}{2}\right) = 1 - \frac{k^2\sigma^2\Delta t}{2} + \frac{1}{2}\left(\frac{k^2\sigma^2\Delta t}{2}\right)^2 + \cdots.$$

The first term tells (reminds) us that

$$p = \frac{\sigma^2\Delta t}{2\Delta x^2},$$

while the second term can be matched (for faster convergence) if

$$p = \frac{1}{6}.$$

The interesting thing about this is that the first four moments (including skew and kurtosis) of the individual steps are as follows:

$$\langle 1 \rangle = p + (1 - 2p) + p = 0$$

$$\langle \Delta x \rangle = p\Delta x - p\Delta x = 0$$

$$\langle \Delta x^2 \rangle = p\Delta x^2 + p\Delta x^2 = \sigma^2\Delta t$$

$$\langle \Delta x^3 \rangle = p\Delta x^3 - p\Delta x^3 = 0$$

$$\langle \Delta x^4 \rangle = p\Delta x^4 + p\Delta x^4 = 2p(\sigma^2\Delta t)^2$$

Skew and kurtosis are defined as

$$skew = \frac{\langle (\Delta x - \langle \Delta x \rangle)^3 \rangle}{\langle (\Delta x - \langle \Delta x \rangle)^2 \rangle^{3/2}}$$

$$kurtosis = \frac{\langle (\Delta x - \langle \Delta x \rangle)^4 \rangle}{\langle (\Delta x - \langle \Delta x \rangle)^2 \rangle^2}$$

and they are 0 and 3 respectively, for a Gaussian distribution. This implies that

$$p = \frac{1}{6}.$$

The von Neumann stability analysis gives the result that faster convergence occurs if we match Gaussian kurtosis (and skew) by adjusting the space-step size depending on the time-steps, such that

$$p_{j+1} = \frac{1}{6}, \quad p_j = \frac{2}{3}, \quad p_{j-1} = \frac{1}{6};$$

$$\Delta x = \sigma\sqrt{3\Delta t}.$$

Returning to the nondrifting grid and looking for a mode of the form $C_{i,j} = \zeta^{\frac{T}{\Delta t}} e^{ikx}$, we find

$$C_{i-1,j} = \exp(-r_s \Delta t)(p_{j+1} C_{i-1,j+1} + p_j C_{i-1,j} + p_{j-1} C_{i-1,j-1}),$$

$$p_{j-1} = \left(-\alpha e^{\Delta x} + e^{\frac{\Delta x}{2}} \beta\right)/\det, \quad p_{j+1} = \left(-\alpha e^{-\Delta x} + e^{\frac{-\Delta x}{2}} \beta\right)/\det,$$

$$p_j = 1 - p_{j+1} - p_{j-1};$$

$$\alpha = 2\sinh(\Delta x)(e^{r_{short}\Delta t} - 1), \quad \beta = 2\sinh\left(\frac{\Delta x}{2}\right)(e^{(\sigma^2 + 2r_{short})\Delta t} - 1),$$

$$\det = 2\sinh(2\Delta x) - 4\sinh(\Delta x)$$

$$\Rightarrow \zeta = \exp(-r_s \Delta t)(p_{j+1} e^{ik\Delta x} + p_j + p_{j-1} e^{-ik\Delta x}) = \exp(-r_s \Delta t)\zeta_0.$$

The full solution shows for one step:

$$f = \exp[-r_s \Delta t] \exp\left[-\frac{\sigma^2 k^2 \Delta t}{2} + ik(r_{short} - \sigma^2/2)\Delta t\right] \exp[ikx]$$

and this *is* the limit for grid steps going to zero.

For finite grid step size, however, we can improve accuracy if we compare the small step expansion of the two results for one time-step. (See Appendix C Expanding the von Neumann Stability Mode for the Discretized Black-Scholes Equation.)

$$\zeta = \exp(-r_s \Delta t)\zeta_0,$$

$$\zeta_0 = 1 + \left(-\frac{\sigma^2 k^2}{2} + ik\left(r_{short} - \frac{\sigma^2}{2}\right)\right)\Delta t$$

$$+ \frac{1}{2}\left(-\frac{\sigma^2 k^2}{2} + ik\left(r_{short} - \frac{\sigma^2}{2}\right)\right)^2 \Delta t^2$$

$$+ \frac{1}{2}\left(\frac{1}{6p} - 1\right)\left\{\left(\frac{\sigma^2 k^2}{2}\right)^2 - i\sigma^2 k^3\left(r_{short} - \frac{\sigma^2}{2}\right)\right.$$

$$\left. + \sigma^2 k^2\left(3r_{short} + \frac{\sigma^2}{4}\right) + i\sigma^2 k\left(2r_{short} + \frac{\sigma^2}{2}\right)\right\}\Delta t^2$$

$$+ O(\Delta t^3).$$

The full solution tells us that this should ideally read as

$$\zeta_0^{t\,\arg et} = 1 + \left(-\frac{\sigma^2 k^2}{2} + ik(r_{short} - \sigma^2/2)\right)\Delta t$$

$$+ \frac{1}{2}\left(-\frac{\sigma^2 k^2}{2} + ik(r_{short} - \sigma^2/2)\right)^2 \Delta t^2 + O(\Delta t^3).$$

Clearly, the same result is obtained,

$$\mu = \frac{1}{6}.$$

So, for a particular number of time-steps (which determines the time-step size), we have a best accuracy space-step size, expressed by

$$\Delta x = \sigma\sqrt{3\Delta t}.$$

As long as the drift is not too great the weights will always be positive:

$$\alpha = 2\sinh(\Delta x)(e^{r_{short}\Delta t} - 1),$$

$$\beta = 2\sinh\left(\frac{\Delta x}{2}\right)(e^{(\sigma^2 + 2r_{short})\Delta t} - 1),$$

$$\det = 2\sinh(2\Delta x) - 4\sinh(\Delta x);$$

$$p_{j-1} = \left(-\alpha e^{\Delta x} + e^{\frac{\Delta x}{2}}\beta\right)/\det,$$

$$p_{j+1} = \left(-\alpha e^{-\Delta x} + e^{\frac{-\Delta x}{2}}\beta\right)/\det,$$

$$p_j = 1 - p_{j+1} - p_{j-1}.$$

If any of the weights are negative, then the selection of the three relevant grid points is not correct due to high drift. In fact we see there is an upper bound on the time-step size:

$$\Delta x > (r - r_S)\Delta t$$

$$\Rightarrow$$

$$\sqrt{3\Delta t\sigma^2} > (r - r_S)\Delta t$$

$$\Rightarrow$$

$$\Delta t < \frac{3\sigma^2}{(r - r_S)}.$$

If this condition is not satisfied—low volatility calculations or high rate environment—then the three points closest to the drifted stock price should be used and the weights rederived. This will give a new nonnegative set of weights.

9.3.3. Call/Put Options: Grid Point Shift Factor for Higher Accuracy

When valuing an option on a grid, for example a grid in x-space where

$$S = Ke^{x-\left(r_S-r_S-\frac{\sigma^2}{2}\right)(T-t)}$$

with a grid in place,

$$x = (\delta + n)\Delta x \quad 0 < \delta < 1,$$

then, due to imposing Gaussian kurtosis on the weights of a trinomial grid (see section 9.3.2 Gaussian Kurtosis (and Skew $= 0$), Faster Convergence), we know the value of Δx satisfies

$$\Delta x^2 = 3\sigma^2 \Delta t.$$

This ensures fast convergence on the trinomial grid for every Fourier component of the payoff, given a particular value of the time-step Δt (see section see section 9.3.2 Gaussian Kurtosis (and Skew $= 0$), Faster Convergence). Furthermore, this ensures Gaussian kurtosis for each step.

However, we may still ask the question, what is the optimal grid positioning around the strike of the option, namely, what is the optimal value for δ?

If we find the Fourier transform of the solution on the grid and compare it to the Fourier transform of the exact solution, then it becomes apparent that a particular value of δ gives more correct behavior for large wavelength Fourier components and thus greater accuracy for the answer. So consider the payoff $f(x)$ and its Fourier transform on a grid,

$$f(x) = K\max(e^x - 1, 0),$$

$$f(k) = \frac{1}{\sqrt{2\pi}} \int_{x=-\infty}^{x=+\infty} \sum_{n=-\infty}^{n=+\infty} \Delta x \delta(x - (\delta + n)\Delta x)e^{-ikx}(e^x - 1)H(x)\,dx$$

where $H(x)$ is the unit step-function; value 0 for $x < 0$ and 1 for $x > 0$. We have inserted a lattice function and thus the inverse Fourier transform will reproduce the values of the function only at the grid points, but its long wavelength, that is, $\Delta x \to 0$ limit will be the same as without the grid. The result of the integral is

$$f(k) = \frac{1}{\sqrt{2\pi}} \left[\frac{\Delta x e^{\delta \Delta x}}{1 - e^{(1-ik)\Delta x}} - \frac{\Delta x}{1 - e^{-ik\Delta x}} \right].$$

In the $\Delta x \to 0$ limit, we get

$$\lim_{\Delta x \to 0} f(k) = \frac{1}{\sqrt{2\pi}} \left[\frac{-1}{(1 - ik)} + \frac{i}{k} \right].$$

This is the Fourier transform of the payoff (without a grid). Now consider the case where $\Delta x > 0$ and expand the Fourier transform on the grid in powers of k, that is,

$$f(k) = \frac{i}{k} + \left(\frac{\Delta x e^{\delta \Delta x}}{1 - e^{\Delta x}} - \frac{\Delta x}{2} + \delta \Delta x \right) + O(k).$$

Now the correct Fourier transform has expansion in k,

$$payoff(k) = \frac{i}{k} - 1 + O(k)$$

and therefore, because the behavior of the Fourier transform at small values of k corresponds to large x behavior of the solution, a more accurate solution results if we choose the grid positioning relative to strike as implicitly determined by choosing Δx to make the second-order terms match,

$$\frac{\Delta x e^{\delta \Delta x}}{1 - e^{\Delta x}} - \frac{\Delta x}{2} + \delta \Delta x = -1.$$

This is the principal result. However, consider the fact that for most grids $\Delta x << 1$ and so we can expand in powers of Δx. We find the following:

$$\delta = \delta_0 + \Delta x \delta_1 + O(x^2),$$

$$\delta_0 = \frac{1}{2} \pm \frac{1}{\sqrt{12}},$$

$$\delta_1 = -\frac{1}{36}.$$

The result for a put gives exactly the same answer for δ_0 but $+\frac{1}{36}$ for δ_1.

9.3.4. Dividends on the Underlying Equity

If a dividend is to be paid for holders of a stock on a future date, this will cause call options on this stock to be of lower value—all other things being equal. One justification for this is that the riskless hedging strategy requires paying the dividend on the short stock to the stock owner. Another view is that any derivative with an embedded stock forward (such as a call option) is less valuable because, even though the holder *does* own the stock at a future (expiration) date, the holder does not own any dividends paid between today and the expiration date.

One way to model dividends is to assume that the date and payment amount (as a currency amount or as a percentage of the stock price) of the dividend are known ahead of time with complete certainty. The stock price is modeled as "gapping" down by the exact amount of the dividend across the dividend ex-date; the first day on which the stock trades without the dividend. The caveat we are forced to note is that if the stock trades at a price less than the dividend, it is assumed it jumps to zero. To model derivatives on this stock, the price curve is shifted by the dividend amount, expressed as

$C(S, t) :$

$$C_{before\ dividend}(S, divXdate) = \begin{cases} C_{after\ dividend}(S(1 - prop) - div, divXdate) : \\ S > \frac{div}{(1-prop)} \\ C_{afterdividend}(0, t = divXdate) : \\ S \leq \frac{div}{(1-prop)} \end{cases}$$

where *prop* is a dividend payment proportional to the stock price, and *div* is a fixed dividend in the currency of the stock. The result is that (going forward in time) as the holder crosses a dividend X-date, the stock price gaps down by the dividend while the derivative does *not* change price.

NOTE ON CALIBRATION

Volatility inputs to these models will be defined as stock price changes without dividend jumps. This is a significant fact because including dividend jumps can lead to much higher volatilities and, most often, the historical volatility calculators used by practitioners *include* dividend jumps.

9.3.5. American Exercise

American exercise means that an option is written where the buyer can exercise at any time from purchase date to expiration date (In contrast to an American exericse, *European* is defined as the holder being able to exercise only at expiration and not before. So far all discussion in this book has been in regard to European exercise). At every instant the option holder has a choice to hold the option or exercise. This will require an analysis with multiple embedded option valuation. Here we discuss call options.

On a grid, the valuation algorithm for call options is as follows: At expiration the price payout function is $\max(S - K, 0)$; then propagate this price distribution backward in time, one time-step; then take the maximum of this result (the "hold" value) with

$$Parity = \max(S - K, 0),$$

then continue this iteratively. This is a rational exercise valuation. The holder maximizes his value by choosing between holding and exercising. An exercise boundary is generated, $S = S_*(t)$. This is called a *free-boundary problem*.

Another formulation is to specify a forced exercise boundary, say, $S = \overline{S}(t)$, where the option price is fixed as parity as a boundary condition. This is like a "call" schedule in convertible bonds or warrants. Then the price may be considered a function of this entire boundary function, that is

$$C(S, t, T; [\overline{S}(t)]).$$

The American (free) boundary is the stationary point of this expression that maximizes the price as a function of each point on the boundary; everywhere below the boundary the solution to the Black-Scholes equation is the American option solution, expressed as

$$\overline{S}(t_i) = S_*(t_i): \quad \frac{\partial C(S, t, T; \overline{S}(t_1), \overline{S}(t_2), \ldots, \overline{S}(t_N))}{\partial \overline{S}(t_i)} = 0$$

for all i. This is written in the shorthand of functional analysis as

$$\overline{S}(t) = S_*(t): \quad \frac{\delta C(S, t, T; [\overline{S}(t)])}{\delta \overline{S}(t)} = 0.$$

Analytical approximations to the result of the American option problem have been worked out; the Whaley formula is one of them.*

Here we note the following: At the boundary, delta is continuous because if it were not we could move the boundary point and increase the price, just below the boundary. Thus, along the boundary, and as the limit from below,

$$C(S,t)|_{S \to S_*(t)_-} - S_*(t) - K,$$

$$\left. \frac{\partial C(S,t)}{\partial S} \right|_{S \to S_*(t)_-} = 1.$$

Everywhere above the boundary,

$$C(S,t)|_{S > S_*(t)} = S - K, \qquad \left. \frac{\partial C(S,t)}{\partial S} \right|_{S > S_*(t)} = 1, \qquad \left. \frac{\partial^2 C(S,t)}{\partial S^2} \right|_{S > S_*(t)} = 0,$$

and so, as the limit from above

$$C(S,t)|_{S \to S_*(t)_+} = S_*(t) - K,$$

$$\left. \frac{\partial C(S,t)}{\partial S} \right|_{S \to S_*(t)_+} = 1, \qquad \left. \frac{\partial^2 C(S,t)}{\partial S^2} \right|_{S \to S_*(t)_+} = 0.$$

Now, moving along the boundary as a limit from below,

$$\frac{dC(S_*(t),t)}{dt} = \left. \frac{\partial C(S,t)}{\partial S} \right|_{S \to S_*(t)_-} \frac{dS_*(t)}{dt} + \left. \frac{\partial C(S,t)}{\partial t} \right|_{S \to S_*(t)_-} = \frac{dS_*(t)}{dt}$$

$$\Rightarrow \left. \frac{\partial C(S,t)}{\partial t} \right|_{S \to S_*(t)_-} = 0$$

Then we may put these values into the Black-Scholes equation to find the gamma limit as we approach the boundary from below,

$$\left[\frac{\partial C(S,t)}{\partial t} + \frac{\sigma^2(S,t)S^2}{2} \frac{\partial^2 C(S,t)}{\partial S^2} + (r_\$ - r_{stock})S \frac{\partial C(S,t)}{\partial S} - r_\$ C(S,t) \right]\Bigg|_{S \to S_*(t)_-} = 0,$$

$$\frac{\sigma^2(S_*,t)S_*^2}{2} \Gamma + (r_\$ - r_{stock})S_* - r_\$(S_* - K) = 0$$

$$\Rightarrow \lim_{S \to S_{*-}} \Gamma = \frac{2(r_{stock}S_* - r_\$ K)}{\sigma^2(S_*,t)S_*^2}.$$

*See G. Barone-Adesi and R. E. Whaley (1987).

Hence gamma is discontinuous across the American exercise boundary (except at maturity), while delta is continuous. This result is useful for numerical evaluations of American options because we find the exercise boundary during the calculation, and this formula means that we know what size of gamma discontinuity to expect. This helps immensely for interpolation algorithms because more than three-point interpolation through discontinuities in the second-order derivative causes unstable results.

Also note that for an American call option the value of $S_*(t)$ for $t = T$ and the steady-state solution ($t \to -\infty$) may be determined. Take the $t \to T$ limit of the Black-Scholes (European) option formula equal to exercise to find the point that is "neutral" to exercise:

$$Se^{-r_{stock}(T-t)} - Ke^{-r_\$(T-t)} = S - K,$$

$$\Rightarrow S - K - r_{stock}\Delta t S + r_\$\Delta t K = S - K + O(\Delta t^2)$$

$$\Rightarrow S_*(T) = K\frac{r_\$}{r_{stock}}.$$

There is only an exercise boundary (and hence exercise) if

$$1 < \frac{r_\$}{r_{stock}} < \infty,$$

Also, a steady-state solution may be determined,

$$t \to -\infty$$

$$\frac{\sigma^2 S^2}{2}\frac{\partial^2 C(S)}{\partial S^2} + (r_\$ - r_{stock})S\frac{\partial C(S)}{\partial S} - r_\$ C(S) = 0: \quad S < S^*(t = -\infty)$$

$$C(S^*) = S^* - K, \quad \left.\frac{\partial C(S)}{\partial S}\right|_{S=S^*} = 1;$$

$$S < S^*: \quad C(S) = AS^\nu \quad \text{where} \quad \frac{\nu(\nu-1)}{2}\sigma^2 + \nu(r_\$ - r_{stock}) - r_\$ = 0$$

$$\Rightarrow AS^{*\nu} = S^* - K, \quad A(\nu-1)S^{*\nu-1} = 1$$

where the last two formulae determine A and S_*, to give

$$S_*(t \to -\infty) = K\frac{(\nu-1)}{(\nu-2)},$$

where ν is determined by $\frac{\sigma^2}{2}\nu^2 + \nu\left(r_\$ - r_{stock} - \frac{\sigma^2}{2}\right) - r_\$ = 0.$

9.3.6. 2-D Models, Correlation and Variable Changes

Given two independent Wiener processes $dz_1(t)$ and $dz_2(t)$, we can construct two arbitrary stock price paths $S_1(t)$ and $S_2(t)$, of different mean returns and volatilities and arbitrary correlation of percentage changes (see section 3.4 Other Processes: Multivariable Correlations).

Defining

$$dx_1 = \alpha_1(t)(dz_1(t) + \beta(t)\, dz_2(t)) + \mu_1(t)\, dt;$$

$$dx_2 = \alpha_2(t)(dz_1(t) - \beta(t)\, dz_2(t)) + \mu_2(t)\, dt,$$

$$dz_1 = \frac{1}{2}\left[\frac{(dx_1 - \mu_1\, dt)}{\alpha_1} + \frac{(dx_2 - \mu_2\, dt)}{\alpha_2}\right],$$

$$dz_2 = \frac{1}{2\beta}\left[\frac{(dx_1 - \mu_1\, dt)}{\alpha_1} - \frac{(dx_2 - \mu_2\, dt)}{\alpha_2}\right];$$

$$\alpha_1(t) = \sigma_1(t)\sqrt{\frac{1+\rho(t)}{2}}, \quad \alpha_2(t) = \sigma_2(t)\sqrt{\frac{1+\rho(t)}{2}}, \quad \beta(t) = \sqrt{\frac{1-\rho(t)}{1+\rho(t)}},$$

and noting that these changes are normally distributed (not lognormally) we may use this as a model of stock price returns,

$$\langle dx_1 \rangle = \mu_1(t)\, dt, \quad \langle (dx_1 - \langle dx_1 \rangle)^2 \rangle = \sigma_1^2(t)\, dt;$$

$$\langle dx_2 \rangle = \mu_2(t)\, dt, \quad \langle (dx_2 - \langle dx_2 \rangle)^2 \rangle = \sigma_2^2(t)\, dt;$$

$$\frac{\langle dx_1\, dx_2 \rangle - \langle dx_1 \rangle \langle dx_2 \rangle}{\sigma_1(t)\sigma_2(t)} = \rho(t)\, dt.$$

Because the path end-points are given by

$$X_1 = \int_{t=0}^{t=T} dx_1, \, X_2 = \int_{t=0}^{t=T} dx_2,$$

we may define the stock prices by

$$S_1 = e^{X_1}, \quad S_2 = e^{X_2}, \quad \tilde{\mu}_1 = \mu_1 + \frac{\sigma_1^2}{2}, \quad \tilde{\mu}_2 = \mu_2 + \frac{\sigma_2^2}{2},$$

which corresponds to a stock price process

$$dS_1 = S_1 \tilde{\mu}_1 \, dt + S_1 \sigma_1 \, dx_1, \quad dS_2 = S_2 \tilde{\mu}_2 \, dt + S_2 \sigma_2 \, dx_2,$$

$$\langle dS_1 \, dS_2 \rangle = S_1 S_2 \rho_{12} \sigma_1 \sigma_2 \, dt.$$

Constructing a riskless portfolio of one derivative that depends on (i.e., has a delta to) both stocks, and a shorted amount (i.e., Δ_1 and Δ_2) of each stock, it can easily be shown—following the steps of section 5.2 Derivation of Black-Scholes Equation, Risk-Neutral Pricing—that any derivative must satisfy

$$\frac{\partial C}{\partial t} + \frac{S_1^2 \sigma_1^2}{2} \frac{\partial^2 C}{\partial S_1^2} + \frac{S_2^2 \sigma_2^2}{2} \frac{\partial^2 C}{\partial S_2^2} + S_1 \sigma_1 S_2 \sigma_2 \rho_{12} \frac{\partial^2 C}{\partial S_1 \partial S_2}$$

$$+ S_1 R_1 \frac{\partial C}{\partial S_1} + S_2 R_2 \frac{\partial C}{\partial S_2} - r_\$ C = 0$$

where $R_1 = r_\$ - r_{S_1}$ and $R_2 = r_\$ - r_{S_2}$ are the short rates for each stock borrow.

This is easier to solve in uncorrelated nondrifting variables, which can be worked out, that is,

$$\frac{\partial C}{\partial t} + \frac{1}{2} \frac{\partial^2 C}{\partial Z_1^2} + \frac{1}{2} \frac{\partial^2 C}{\partial Z_2^2} - r_\$ C = 0,$$

$$Z_1 = \int_{t'=0}^{t'=t} dz_1(t'), \quad Z_2 = \int_{t'=0}^{t'=t} dz_2(t').$$

Thus the derivative's payout function must be specified and then translated into the Z-variables and then the solution to the above equation analytically or numerically evaluated.

To numerically evaluate on a 3-D grid over t, Z_1 and Z_2 we need to specify the following moments:

$$C_{00} = K e^{-r_\$ (T-t)},$$

$$C_{10} = S_1 e^{-r_1 (T-t)},$$

$$C_{01} = S_2 e^{-r_2 (T-t)},$$

$$C_{20} = S_1^2 e^{-r_1 (T-t)} e^{(\sigma_1^2 + 2R_1)(T-t)},$$

$$C_{02} = S_2^2 e^{-r_1 (T-t)} e^{(\sigma_2^2 + 2R_2)(T-t)},$$

$$C_{11} = S_1 S_2 e^{-r_1 (T-t)} e^{(\rho \sigma_1 \sigma_2 + R_1 + R_2)(T-t)}.$$

The last moment is redundant because the correlation is ensured by the variable change, and so we need to ensure that only the five moments,

$$C_{00} = Ke^{-r_\$(T-t)},$$

$$C_{10} = S_1 e^{-r_1(T-t)},$$

$$C_{01} = S_2 e^{-r_2(T-t)},$$

$$C_{20} = S_1^2 e^{-r_1(T-t)} e^{(\sigma_1^2 + 2R_1)(T-t)},$$

$$C_{02} = S_2^2 e^{-r_1(T-t)} e^{(\sigma_2^2 + 2R_2)(T-t)}$$

are valued exactly.

For example, if correlation and volatilities are constant (not functions of time), we may use the variable changes,

$$S_1 = \exp(\alpha_1 Z_1 + \alpha_1 \beta Z_2 - R_1(T-t)),$$

$$S_2 = \exp(\alpha_2 Z_1 - \alpha_2 \beta Z_2 - R_2(T-t)),$$

where

$$\alpha_1 = \sigma_1 \sqrt{\frac{1+\rho}{2}}, \quad \alpha_1 \beta = \sigma_1 \sqrt{\frac{1-\rho}{2}},$$

$$\alpha_2 = \sigma_2 \sqrt{\frac{1+\rho}{2}}, \quad \alpha_2 \beta = \sigma_2 \sqrt{\frac{1-\rho}{2}}.$$

Then we put the equation on a grid,

$$Z_1 = z_1 \Delta z_1, \quad Z_2 = z_2 \Delta z_2,$$

where z_1, z_2 are integers and Δz_1, Δz_2 are the step sizes. Thus to get a new price value at $t-1$, we need the five nearest points (which form a cross on the 2-D z_1, z_2 plane), expressed as

$$C_{t-1,z_1,z_2} = e^{-r_\$ \Delta t}$$

$$(p_{-0} C_{t,z_1-1,z_2} + p_{0-} C_{t,z_1,z_2-1} + p_{00} C_{t,z_1,z_2} + p_{+0} C_{t,z_1+1,z_2} + p_{0+} C_{t,z_1,z_2+1}).$$

The five weights are then determined by ensuring that the five moments are exactly solved:

$$1 = p_{-0} + p_{0-} + p_{00} + p_{+0} + p_{0+},$$

$$e^{-R_1 \Delta t} = p_{-0} e^{-(\alpha_1 \Delta z_1 + \alpha_1 \beta \Delta z_2 - R_1 \Delta t)} + p_{0-} + p_{00}$$
$$+ p_{+0} e^{(\alpha_1 \Delta z_1 + \alpha_1 \beta \Delta z_2 - R_1 \Delta t)} + p_{0+},$$

$$e^{-R_2 \Delta t} = p_{-0} + p_{0-} e^{-(\alpha_2 \Delta z_1 - \alpha_2 \beta \Delta z_2 - R_2 \Delta t)} + p_{00} + p_{+0}$$
$$+ p_{0+} e^{(\alpha_2 \Delta z_1 - \alpha_2 \beta \Delta z_2 - R_2 \Delta t)},$$

$$e^{(\sigma_1^2 + 2R_1) \Delta t} = p_{-0} e^{-2(\alpha_2 \Delta z_1 - \alpha_2 \beta \Delta z_2 - R_2 \Delta t)} + p_{0-} + p_{00}$$
$$+ p_{+0} e^{2(\alpha_2 \Delta z_1 - \alpha_2 \beta \Delta z_2 - R_2 \Delta t)} + p_{0+},$$

$$e^{(\sigma_2^2 + 2R_2) \Delta t} = p_{-0} + p_{0-} e^{-2(\alpha_2 \Delta z_1 - \alpha_2 \beta \Delta z_2 - R_2 \Delta t)} + p_{00}$$
$$+ p_{+0} + p_{0+} e^{2(\alpha_2 \Delta z_1 - \alpha_2 \beta \Delta z_2 - R_2 \Delta t)}.$$

Not surprisingly, faster convergence results from imposing Gaussian skew and kurtosis in each independent uncorrelated variable separately, so that

$$\Delta z_1 = \sqrt{3 \Delta t}, \quad \Delta z_2 = \sqrt{3 \Delta t}.$$

This is the basic method for putting multifactor models onto a grid because generalizing to three or more variables is straightforward.

For example, the algorithm is straightforward for valuing:

1. A call on two stocks, meaning that the owner has the right but not the obligation to buy two stocks simultaneously for a set price or strike K; the payout is $\max(S_1 + S_2 - K, 0)$ and this is the terminal distribution and then the above weights are used to diffuse backward, with account taken for dividends on the stocks if any.
2. A call on the better of two stocks, meaning that the payout is $\max(\max(S_1, S_2) - K, 0)$: with this terminal distribution the algorithm implementation is otherwise exactly the same.

CHAPTER 10

Credit Spreads

10.1. CREDIT DEFAULT SWAPS (CDS) AND THE CONTINUOUS CDS CURVE

Credit default swap (CDS) contracts are generally contracts in which the buyer agrees to pay Φ_i at the end of every period i, quarterly for example, in return for insurance against default—where a *default event* is defined by the contract as failing to pay coupons or certain specified escrow balances dropping below predefined thresholds, and the like. When default occurs, the buyer may deliver a bond that is specified by the contract and get par in return. If the bonds trade at price R, the *recovery* value, at this time then the CDS contract is worth $1 - R$ in default, as shown in Table 10.1.

Equating the short value and long value at each successive payment date in order to make the at-the-market swap fair value, that is, zero,

$$(1 - p_1) = \frac{1}{1 + \left[\frac{\Phi_1}{(1-R_1)}\right]},$$

$$(1 - p_2) = \frac{1}{1 + \left[\frac{\Phi_2}{(1-R_2)}\right]}.$$

Now consider a continuous payout, $\phi(t)$, which means paying at the end of every day, or instant, instead of every quarter. The survival probability $S_T(t)$, from time t to T, is given by

$$S_T(t) = \exp\left[-\int_{t'=t}^{t'=T}\left(\frac{\phi(t')}{1 - R(t')}\right) dt'\right].$$

TABLE 10.1 Cash flows of a credit default swap I (discrete)

	No Default		Default	
i	Probability	Buyer Is Short Value	Probability	Buyer Is Long Value
1	$(1 - p_1)$	$-\Phi_1(1 - p_1)$	p_1	$p_1(1 - R_1)$
2	$(1 - p_1)(1 - p_2)$	$-\Phi_2(1 - p_1)(1 - p_2)$	$(1 - p_1)p_2$	$(1 - p_1)p_2(1 - R_2)$

Another way to view this is, given a continuous payout curve $\phi(t)$, and a recovery curve $R(t)$, we can implement a variable change $\{\phi(t), R(t)\} \rightarrow \{s(t), R(t)\}$ as

$$s(t) = S_t(0) = \exp\left[-\int_{t'=0}^{t'=t} \left(\frac{\phi(t')}{1 - R(t')} \right) dt' \right],$$

where $s(t)$ is the survival probability from time $t = 0$ to time t. Note that

$$-\frac{ds(t)}{dt} = \frac{\phi(t)}{1 - R(t)} s(t).$$

The negative rate of change of the survival probability is the product of the default probability rate at time t, and the survival probability to time t.

CDS *unwind* or *tear-up value* is the premium paid or received to get out of a CDS contract. The buyer has agreed to pay a fixed amount per quarter for insurance against loss in default. But at a later time, before the expiration of the contract, the buyer decides he wants to unwind the contract and the premium to pay or receive to do this must be calculated. The calculation is as follows: For a continuous spread (payment) curve $\varphi(t)$ and for specified discrete payments in the marketplace now and at set up it is $\{\Phi_i\}, \{\Phi_{set-up_i}\}$. The unwind is the difference between owning and selling, present value using rates and survival probabilities and is expressed as

Unwind value

$$= \sum_i (\Phi_i - \Phi_{set-up_i}) \exp\left[-\int_{u=0}^{u=t_i} \left(\frac{\phi(u)}{1 - R(u)} \right) du \right] \exp\left[-\int_{u=0}^{u=t_i} (r(u)) du \right].$$

The only issue here is how to relate the current market price of the CDS $\{\Phi_i, R_i\}$ to the continuous payout curve $\phi(t)$ and $R(t)$. They are the same curves, merely expressed in a different quote convention. This is much like the fact that a curve of continuous, instantaneous forward interest rates corresponds to a unique curve of interest rate swap prices. Either one can be calculated from the other because the two curves represent the same interest rate environment. The only difference is the quote convention. Consider a period from t to T, where there is a constant danger of default, albeit with recovery $R(t)$, and at the end of the period a payment is made to compensate the issuer for his losses as shown in Table 10.2.

Now we need to add up all the payout values and set this equal to the payment amount as $N \to \infty$, expressed as

$$\sum_{j=1}^{j=N} \Phi \Delta T \exp\left[-\int_0^{t_j} \left(r(s) + \frac{\phi(s)}{1 - R(s)}\right) ds\right]$$

$$= \sum_i \frac{\phi_i}{r_i + \frac{\phi_i}{(1-R_i)}} \left(1 - \exp\left[-\left(r_i + \frac{\phi_i}{1 - R_i}\right)\Delta t_i\right]\right)$$

$$\times \exp\left[-\int_0^{t_{i-1}} \left(r(s) + \frac{\phi(s)}{1 - R(s)}\right) ds\right]$$

where $\Delta T = \frac{T}{N}$, for swap maturity T (years), N payments, and each payment is $\frac{\Phi}{N}$. Note that i runs over all days that the contract is valid and j runs over all the N payments.

In continuous terms this is

$$\Phi \frac{T}{N} \sum_{j=1}^{j=N} \exp\left[-\int_0^{t_j} \left(r(s) + \frac{\phi(s)}{1 - R(s)}\right) ds\right]$$

$$= \int_0^T \phi(t) \exp\left[-\int_0^t \left(r(u) + \frac{\phi(u)}{1 - R(u)}\right) du\right] dt.$$

This formula is a CDS unwind from discrete to continuous payments. This makes sense because the fair, at-the-market value of a swap

TABLE 10.2 Cash flows of a credit default swap II (continued)

	No Default Survival Probability	We Pay	Default Default Probability	Payout
1	$\left(1 - \frac{\phi_1 \Delta t}{1-R_1}\right)$	0	$\frac{\phi_1 \Delta t}{1-R_1}$	$\left(\frac{\phi_1 \Delta t}{1-R_1}\right)(1-R_1) = \phi_1 \Delta t$
2	$\left(1 - \frac{\phi_1 \Delta t}{1-R_1}\right)\left(1 - \frac{\phi_2 \Delta t}{1-R_2}\right)$	0	$\frac{\phi_2 \Delta t}{1-R_2}\left(1 - \frac{\phi_1 \Delta t}{1-R_1}\right)$	$\phi_2 \Delta t\left(1 - \frac{\phi_1 \Delta t}{1-R_1}\right)$
$\vdots$	$\vdots$	$\vdots$	$\vdots$	$\vdots$
N-1	$\prod_{i=1}^{i=N-1}\left(1 - \frac{\phi_i \Delta t}{1-R_i}\right)$	0	$\frac{\phi_{N-1}\Delta t}{1-R_{N-1}}\prod_{i=1}^{i=N-2}\left(1 - \frac{\phi_i \Delta t}{1-R_i}\right)$	$\phi_{N-1}\Delta t\prod_{i=1}^{i=N-2}\left(1 - \frac{\phi_i \Delta t}{1-R_i}\right)$
N	$\sim\left(1 - \frac{\phi \Delta t}{1-R}\right)^{\frac{T}{\Delta t}} = p_T$	$-\Phi p_T$		$\sim \phi_N \Delta t\left(1 - \frac{\phi_i \Delta t}{1-R_i}\right)^{\frac{T}{\Delta t}-1}$

is zero and the only difference is the discrete number of cash flows, $\Phi \frac{T}{N}$, on the left and the infinite number of cash flows, $\phi(t)\,dt$, on the right.

The other viewpoint is that the right-hand side is the value of the possible payout of the CDS. Default between times t and $t + dt$ occurs with probability $p(t)\,dt$, where

$$p(t)\,dt = \frac{\phi(t)}{1 - R(t)}\,dt.$$

This happens only if the CDS survives to time t, which occurs with probability

$$s(t) = \exp\left[-\int_{u=0}^{u=t} \frac{\phi(u)}{1 - R(u)}\,du \right].$$

The typical CDS contract specifies that we get a dollar amount specifying the size of the contract—called the contract *notional* or, for a single bond, *par*—if we deliver the same notional of bond trading in default at or at least somewhere near recovery value R. The market price of the bond in default is *notional times R*. Therefore the value of the CDS contract after a default (and for simplicity for a single bond of par = 1) is

$$1 - R(t),$$

to which we need to apply a discount factor

$$\exp\left[-\int_{u=0}^{u=t} r(u)\,du \right]$$

and then the product of these four factors integrated over t gives the (probability weighted) present value of payouts in default.

For a given CDS curve $\{\Phi(t), R(t)\}$ we will have to use the ladder method to obtain the continuous payout curve $\{\phi(t), R(t)\}$.

10.2. VALUING BONDS USING THE CONTINUOUS CDS CURVE

To correctly value a bond, the calculation requires similar considerations to above. The bond fair value is equal to the sum of the riskless present values of the cash flows multiplied by their survival probabilities plus the evaluation of a continuous probability of default that pays the recovery value $R(t)$.

The fair price of a bond is therefore

$$B_T(t)$$

$$= \sum_{i=1}^{N} c_i \exp \left[- \int_{s=t}^{s=t_i} \left(r(s) + \frac{\phi(s)}{1 - R(s)} \right) ds \right]$$

$$+ \int_{s=t}^{s=T} \frac{\phi(s)R(s)}{1 - R(s)} \exp \left[- \int_{u=t}^{u=s} \left(r(u) + \frac{\phi(u)}{1 - R(u)} \right) du \right] ds.$$

10.3. EQUATIONS OF MOTION FOR BONDS AND CREDIT DEFAULT SWAPS

Focusing on a nonstochastic rates environment for the moment, there is an equation of motion for bonds with a probability of default. The price of such bonds, $B_T(t)$, satisfies by differentiating the formula for a bond introduced in the previous section, that is,

$$\frac{\partial B_T(t)}{\partial t} - \left(r(t) + \frac{\phi(t)}{1 - R(t)} \right) B_T(t) = -\frac{\phi(t)R(t)}{1 - R(t)}.$$

To derive this equation and, in the process, see the optimal hedging strategy for CDS against bonds, we construct a portfolio of one bond and α notional of a pure at-the-market (i.e., zero value) CDS:

$$\Pi = B_T(t) + \alpha CDS,$$

In default : $\quad d\Pi = R(t) - B_T(t) + \alpha(1 - R(t)),$

Nondefault : $d\Pi = \left(\dfrac{\partial B_T(t)}{\partial t} - \alpha\phi(t) \right) dt.$

In default, the bond trades at price R—and therefore value par times R, which, given a bond par of 1, is just R)—while the CDS contract requires that we purchase the same notional of bonds, at price R, and deliver them under the terms of the CDS contract to get back the actual notional. If default does not occur, the bond changes in price by theta and we pay the instantaneous premium for the CDS. This implies that if we hedge with

$$\alpha = \frac{B_T(t) - R(t)}{1 - R(t)}$$

notional of CDS, then we have a riskless portfolio, and so the *non-default* scenario will accrue at the riskless rate, thus

$$\Pi = B_T(t) + \alpha CDS,$$

$$\text{Nondefault}: d\Pi = r(t)\Pi\, dt$$

$$\Rightarrow \frac{\partial B_T(t)}{\partial t} - \alpha\phi(t) = r(t)B_T(t)$$

Rearranging we get

$$\frac{\partial B_T(t)}{\partial t} - \left(r(t) + \frac{\phi(t)}{1 - R(t)}\right)B_T(t) = -\frac{\phi(t)R(t)}{1 - R(t)}.$$

Clearly the optimal hedge is to hedge the cash value of the position with enough CDS notional that the payout in default covers the loss in the bond (rather than, say, a one-to-one CDS notional versus bond notional). This is then consistent with the probabilistic interpretation of the CDS continuous curve, $\phi(t)$.

Note that CDS satisfy a slightly different equation of motion

$$\frac{\partial CDS_T(t)}{\partial t} - \left(r(t) + \frac{\phi(t)}{1 - R(t)}\right)CDS_T(t) = -\phi(t).$$

Thus CDS and bonds satisfy similar pricing equations, but with different source terms due to default. The source term reflects the fact that the bonds pay out recovery R, with instantaneous probability (of default) rate

$$p(t) = \frac{\phi(t)}{1 - R(t)},$$

while CDS pay out $1-R$ with the same probability rate.

CHAPTER 11

Specific Models

11.1. STOCHASTIC RATES AND DEFAULT

The combination of the static default model described in Chapter 10 and a stochastic rates model like the Hull-White discussed in Chapter 8 is now straightforward. The spot rate r—that is, instantaneous riskless borrow rate—follows a risk-neutral process consistent with arbitrage-free bond pricing in the HJM framework,

$$dr = (v^2(t) - k(r - f(t)) + f'(t)) \, dt - \sigma \, dz(t)$$

for an initial (riskless) forward curve $f(t)$; model specifications σ and k (both constants; the spot rate standard deviation and the bond volatility/mean reversion decay rate respectively) and

$$v^2(t) = \sigma^2 \frac{(1 - e^{-2kt})}{2k}.$$

According to this model, a riskless zero-coupon bond, $Z_T(r, t)$, has value

$$Z_T(r, t) = \exp\left(-a_T(t)(r - f(t)) - b_T(t) - \int_t^T f(s) \, ds\right)$$

where

$$a_T(t) = \frac{1}{k}(1 - e^{-k(T-t)})$$

$$b_T(t) = \frac{\sigma_0^2}{4k} a_T^2(t)(1 - e^{-2kt})$$

and it satisfies the pricing equation implied by the short rate process,

$$\frac{\partial Z_T(r, t)}{\partial t} + \frac{\sigma^2}{2} \frac{\partial^2 Z_T(r, t)}{\partial r^2} + (v^2(t) - k(r - f(t))$$

$$+ f'(t)) \frac{\partial Z_T(r, t)}{\partial r} - r Z_T(r, t) = 0.$$

Now a simple model combining rates and default might be: the spot rate continues the above process independently of default; default is a *jump-probability* event (i.e., it is not stochastic *and* a known constant probability curve for default exists) and the probability of default is not correlated to the riskless rates curve. The arbitrage-free value of a zero under these restrictions is the sum of two terms. The first term is the riskless zero for par redemption multiplied by survival probability. The second term is the survival probability to time s, multiplied by the default probability at time s, multiplied by the riskless value of recovery R at time s, and added up over all times s from today t to maturity T, expressed as

$$B_T(r,t) = Z_T(r,t) \exp\left[-\int_{s=t}^{s=T} \left(\frac{\phi(s)}{1-R(s)}\right) ds\right]$$
$$+ \int_{s=t}^{s=T} \frac{\phi(s)R(s)}{1-R(s)} Z_s(r,t) \exp\left[-\int_{u=t}^{u=s} \left(\frac{\phi(u)}{1-R(u)}\right) du\right] ds.$$

It follows that this value solves the pricing equation,

$$\frac{\partial B_T}{\partial t} + \frac{\sigma^2}{2} \frac{\partial^2 B_T}{\partial r^2} + (v^2(t) - k(r - f(t)) + f'(t)) \frac{\partial B_t}{\partial r}$$
$$- \left(r(t) + \frac{\phi(t)}{1-R(t)}\right) B_t = \frac{\phi(t)R(t)}{1-R(t)}.$$

From what we have developed so far in this book, we know the theoretical dynamic replication strategy. That is, hedging a default risk zero-coupon bond with a *default*-riskless (think Treasury) zero and some notional of credit default swap (CDS) under which this bond is the only deliverable. CDS contracts often have a whole list of possible deliverable bonds after some defined technical default event and this raises the issue of the value of the optionality of the cheapest to deliver, which we ignore here.

For recovery, $R = 0$, the theoretically ideal amounts of the two hedge instruments are a riskless zero-coupon bond of same present value PV (i.e., less face value of the treasury than the zero bond to be hedged) and CDS notional equal to the present value. These two amounts change as we go through time. Therefore the hedging strategy is dynamic. The costs of the two hedging instruments may be entered into the formula for the (risky) zero-coupon bond and, given that all of the above assumptions hold, an arbitrage may be executed.

For recovery not zero it is a little more complicated. We require the riskless zeros of every day's maturity to match the recovery payout, in an amount to match their PV, plus an amount of CDS to match the loss in the bond in default $(PV - R)$ with its own profit, $1 - R$, namely, CDS notional of amount (bond price-recovery)/(1—recovery) per bond.

Developments to this are apparent. Obviously we already have a universe of possibilities for modeling—two-driver stochastic default probabilities and stochastic riskless rates—that enable a deeper understanding of bonds versus credit default swap trading and rates hedging and the like. Generally new drivers introduce at least one or two more calibration numbers to fit the model to the markets. The basic problem is information input versus output. The simplest model will have only one input number (a one-driver volatility for example) and almost certainly not fit the market for most cases. Then the most complex model will have as many inputs as all pieces of information in the marketplace and will not be testable until a new event occurs. It will almost certainly be wrong because it is untested. This is the basic problem of financial modeling: the trade-off between meaningful predictive modeling and calibration requirements.

The markets are not systems governed by static rules (or even quasi-static, so-called *adiabatic* laws in physics) and this is the essence of the *art* of financial modeling. The real use of modeling is as a frame of reference, or paradigm, in which to look at prices. To compare prices when markets are acting like the simplest models and when they are not adds tools to the trader's and risk manager's toolbox. Clearly, the simplest models, such as the Black-Scholes option formula, will never go away whether or not these models are the best available for arbitrage trading or risk management.

11.2. CONVERTIBLE BONDS

A convertible bond (CB) has all the specifications of a straight bond; a coupon schedule, maturity, perhaps puts and calls. In addition, it is also convertible at the holder's option—American style; at any time—into a fixed number of shares called the conversion ratio (CR) of a certain stock. If the issuer offers conversion into a stock other than its own, the bonds are known as *exchangeable*.

There are many ways to model CBs. Here we assume a probability curve for default and recovery as outlined earlier. This ensures that a market CDS curve can be used to price convertibles. We assume a lognormal distribution of stock price changes, with a term structure for rates, volatility, and equity. A simple pricing equation is

$$\frac{\partial CB}{\partial t} + \frac{S^2 \sigma^2}{2} \frac{\partial^2 CB}{\partial S^2} + S \left(r_\$ + \frac{\phi(t)}{1 - R(t)} - r_S \right) \frac{\partial CB}{\partial S}$$

$$- \left(r_\$(t) + \frac{\phi(t)}{1 - R(t)} \right) CB(t) = -\frac{\phi(t)R(t)}{1 - R(t)}.$$

However, to understand this model and how to implement it, the following bottom-up development of the model is far more transparent.

First, we assume CDS trade, and for which the convertible would be deliverable after a default event occurs. We also assume that in default we *always* recover $1 - R(t)$. This is actually not a trivial assumption because the convertible bond could conceivably be worth more in default because it is convertible, that is, in the case a technical default event has occurred but the stock is still high. Clearly, exchangeables will generally not be described by this model. But it simplifies the analysis considerably to make the assumption that this never happens. Thus the CDS market gives us curves $\phi(t), R(t)$.

We begin by assuming that the recovery value is zero. The convertible bond can be best understood using the first three moments of the distribution implicit in the pricing formula on the previous page. If the CB is not convertible then we want to get the same result as for straight bonds, which is a sum of cash flows, and this implies:

$$\frac{\partial CB(t)}{\partial t} - \left(r_\$(t) + \frac{\phi(t)}{1 - R(t)} \right) CB(t) = 0.$$

If the CB has no cash flows and mandatory conversion at maturity, then we want the result to be a pure stock forward. This implies that

$$CB = Se^{-r_\$(T-t)}$$

is a solution and this therefore fixes the drift term. Thus the equation of motion for a convertible bond (without recovery) is

$$\frac{\partial CB}{\partial t} + \frac{S^2 \sigma^2(t)}{2} \frac{\partial^2 CB}{\partial S^2} + S \left(r_\$(t) + \frac{\phi(t)}{1 - R(t)} - r_\$(t) \right) \frac{\partial CB}{\partial S}$$

$$- \left(r_\$(t) + \frac{\phi(t)}{1 - R(t)} \right) CB = 0.$$

The curvature term is best understood by the following argument. Make two variable changes, the first to discount the CB price function and the second to shift the stock price by this same drift:

$$CB' = CB \exp \left(- \int_{t=0}^{t=t_i} \frac{\phi(t)}{1 - R(t)} \, dt \right),$$

$$S' = S \exp \left(- \int_{t=0}^{t=t_i} \frac{\phi(t)}{1 - R(t)} \, dt \right).$$

With these new variables (dropping the primes) the equation of motion is

$$\frac{\partial CB}{\partial t} + \frac{S^2 \sigma^2(t)}{2} \frac{\partial^2 CB}{\partial S^2} + S(r_\$(t) - r_S(t)) \frac{\partial CB}{\partial S} - r_\$(t)CB = 0$$

and the boundary conditions are implemented very simply: discount all cash flows by the spread to riskless rate prior to calculation (but not the riskless rate $r_\$(t)$, itself), that is, assume the cash flows c_i, in the convertible are replaced by

$$c_i' = c_i \exp\left(-\int_{t=0}^{t=t_i} \frac{\phi(t)}{1 - R(t)} dt\right)$$

and equity boundary conditions such as $parity = S \times CR$ are preserved.

This equation of motion has a very natural interpretation: it is the optionality embedded in the convertible expressed in riskless currency. The coupons need to be changed from risky, or *corporate*, dollars into riskless dollars via the *forward exchange* rate determined by the CDS curve (note that this will naturally shift the effective strike or forward conversion price of the CB down). The resulting calculation is the usual riskless option valuation of Black-Scholes, and this justifies the curvature term as written.

The last, and very important issue, is adding in the default value associated with recovery, to get the full value of the CB. We can do this simply by noting that the implementation of a numerical algorithm for solving the equation of motion will include a diffusion algorithm over finite-sized time-steps, Δt say. We can add the recovery value between t and $t + \Delta t$ (and assuming curves are flat over the time-steps):

Recovery$(t \mapsto t + \Delta t)$

$$= \int_{s=t}^{s=t+\Delta t} \frac{\phi(s)R(s)}{1 - R(s)} \exp\left[-\int_{u=0}^{u=s}\left(r(u) + \frac{\phi(u)}{1 - R(u)}\right) du\right] ds$$

$$= \exp\left[-\int_{u=0}^{u=t}\left(r(s) + \frac{\phi(s)}{1 - R(s)}\right) ds\right]\left(\frac{\phi(t)R(t)}{1 - R(t)}\right)$$

$$\times \left(\frac{1 - \exp\left[-\left(r(t) + \frac{\phi(t)}{1-R(t)}\right)\Delta t\right]}{\left(r(t) + \frac{\phi(t)}{1-R(t)}\right)}\right)$$

and noting that in the above variable choice we require riskless cash values to be added (because the numerical algorithm itself accounts only for risk-free discounting), and thus the recovery term to be added at each time-step is

$$\exp\left[-\int_{u=0}^{u=t}\frac{\phi(s)}{1-R(s)}ds\right]\left(\frac{\phi(t)R(t)}{1-R(t)}\right)\left(\frac{1-\exp\left[-\left(r(t)+\frac{\phi(t)}{1-R(t)}\right)\Delta t\right]}{\left(r(t)+\frac{\phi(t)}{1-R(t)}\right)}\right).$$

In summary, the algorithm for CB valuation is set forth in the following steps using *rational exercise* assumptions:

1. Convert all coupons, puts and par into riskless currency from corporate currency $c_i' = c_i \exp\left(-\int_{t=0}^{t=t_i}\frac{\phi(t)}{1-R(t)}dt\right)$.
2. Value the convertible on a 5-standard-deviation-wide grid using the methods outlined in section 9.3 solving the Black-Scholes Equation Numerically. The nondrifting grid weights are the simpler choice for implementation because the payouts never need to be transformed into different variables.
3. At maturity maximize riskless par with parity, the conversion value, $S \times CR$, where CR is the conversion ratio of the bond, to generate the initial price slide vector.
4. Diffuse, and risklessly discount backward in time using the weights from earlier plus the recovery value for the time-step, that is

$$CB_{t,x} = \exp(-r_t\Delta t)(w_{x-1}CB_{t+\Delta t,x-1} + w_x CB_{t+\Delta t,x}$$

$$+ w_{x+1}CB_{t+\Delta t,x+1}) + \text{Recovery},$$

$$\text{Recovery} = \exp\left[-\int_{u=0}^{u=t}\frac{\phi(s)}{1-R(s)}ds\right]\left(\frac{\phi(t)R(t)}{1-R(t)}\right)$$

$$\times\left(\frac{1-\exp\left[-\left(r(t)+\frac{\phi(t)}{1-R(t)}\right)\Delta t\right]}{\left(r(t)+\frac{\phi(t)}{1-R(t)}\right)}\right).$$

5. For American-style exercise, maximize the result with parity.
6. If the CB is putable, then maximize the price with the (riskless) put value on the relevant date(s).
7. If the bond is also callable, then *minimize* the value with the call value on the call date. *Call value* is itself the maximum of conversion after call and call price because the typical CB spec allows conversion after call. In fact, the more precise valuation takes into account the optionality of the notice period. Commonly, a CB call spec is the following: the issuer gives notice of call according to a predefined

schedule of call prices and notice period. Then during the notice period the holder may convert or collect the offered cash at the end of the notice period. This correct call value is the max of call price and conversion after call plus a notice period American option value. The modeler can decide to switch the calls into riskless currency using the default probability curve or assume, because the holder is *short* these calls (i.e., they are issuer options), that they are already in riskless currency and so decide not to switch them.

This latter technique gives a slightly higher theoretical value for CBs around call and is not unreasonable because calls are often not exercised perfectly rational in the marketplace. That is, issuers often do not call bonds as soon as it makes financial sense to do so—perhaps because they would rather give out stock than cash. Thus they let the stock run a little higher before calling and then are more confident that the investors will choose conversion, and not cash, after call. Many practitioners even put a "call bump" into the model to reflect this (i.e., increase all call prices by 10 percent or 15 percent).

8. Model any provisional calls. This is a situation in which call is allowed to occur only if the stock is above a given trigger. To model this: minimize the hold value with the call only above the trigger and leave unchanged below. It can lead to small regions of negative delta, which reflects the holder's potential increase in value if the stock drops below the trigger and call by the issuer is no longer allowed.

9. Apply an algorithm to include dividends such as the one outlined in section 9.3.4 Dividends on the Underlying Equity. Note that the decision about whether to discount dividends at the riskless rate or risky rate needs to be taken.

10. Add the riskless value of any coupons paid to the CB. The riskless value is the coupon cash values discounted using just the credit spread from valuation date to coupon date (the algorithm ensures further discounting to reflect discounting due to interest rates).

11. Return to step 4 and repeat steps 4 through 10 until the valuation date is reached.

12. Interpolate to obtain theoretical values at a given stock price.

This model is far from the final word on modeling convertible bonds! But it is a very good reference point for the current convertible market because it has term-structured volatility, rates, and credit—and it is consistent with the recovery values of the CDS market, which is being traded simultaneously by many CB traders as of writing.

But even this model, it is fair to say, has problems. Deep in the money, the credit dependence is opposite to a bond. As credit worsens, the CB becomes more valuable due to the recovery value increasing. For a straight bond, this is compensated by the decrease in value of the future cash flows as credit worsens, giving an overall credit dependence that reduces the value of the bond as credit worsens. It is always the user's choice to set recovery to zero to fix this. CDS, however, rarely trade with an agreed implicit recovery of zero.

Another immediate limitation is that any volatility smile is not incorporated (variation of volatility by strike or nonlognormal effects), including dilution effects on the stock because the company is issuing possibly large new amounts of stock at high stock prices and none at lower stock prices.

In summary, it is not clear exactly how the valuation of the CDS market and CB market interact. It would be a safe bet that it is not precisely according to a simple model like this one!.

11.3. INDEX OPTIONS VERSUS SINGLE NAME OPTIONS: TRADING EQUITY CORRELATION

We can use the above algorithm for an n-factor model (see section 9.3.6. 2-D models, correlating and variable changes) to value index options with n components. However, for an index with tens of components or more this can be prohibitively slow.

A good approximation is the following. We value options on the index as if the index was lognormally distributed using the Black-Scholes equation. We need to choose the volatility input to the model to ensure that the theoretical value (TV) of the option agrees as closely as possible with the correct answer. The way to do this is to consider the first moment and second moments of the actual index distribution, assuming that the stocks are lognormally distributed with different volatilities and stock borrows and a correlation, expressed as

$$dS_1 = S_1\mu_1\, dt + S_1\sigma_1 dx_1,$$

$$dS_2 = S_2\mu_2\, dt + S_2\sigma_2 dx_2,$$

$$\langle dS_1 dS_2 \rangle = S_1 S_2 \rho_{12}\sigma_1\sigma_2\, dt.$$

This means that the index distribution cannot be lognormal. But if we match up the first few moments for a lognormal index distribution, we will obtain answers that are a good approximation for derivatives that have strong dependence on these moments and weaker dependence on higher moments. Call and put options depend strongly on the first moment (the mean) and the second moment (the volatility), and more weakly on higher moments. The practitioner must be very careful if using this method to value combinations of long and short options (such as butterflies, etc.), which depend strongly on higher moments.

BUTTERFLIES

A *butterfly* is a combination of puts and calls that has a total terminal distribution that is zero everywhere except for between two strikes, where it is shaped like a triangle. A long call at the lower strike, short two calls struck at halfway between the two strikes, and a long call at the upper strike produce a butterfly.

Returning to the case of modeling a process for the underlying of options that are more straightforward, such as calls or puts that have much weaker dependence on higher moments of the terminal distribution, we may express the value of a two stock index I and its process as

$$I = S_1 + S_2,$$
$$\Rightarrow dI = dS_1 + dS_2$$
$$\Rightarrow I(r_\$ - r_I)\, dt + I\sigma(S_1, S_2)\, dz = S_1(r_\$ - r_1)\, dt + S_1\sigma_1 dz_1$$
$$+ S_2(r_\$ - r_2)\, dt + S_2\sigma_2 dz_2.$$

We want to approximately match to a lognormal index distribution,

$$S_1(r_\$ - r_1)\, dt + S_1\sigma_1 dz_1 + S_2(r_\$ - r_2)\, dt + S_2\sigma_2 dz_2 \cong I(r_\$ - r_I)\, dt + I\sigma_I\, dz.$$

The mean of this match-up can be fixed by setting the first moments to be equal

$$-r_1 S_1\, dt - r_2 S_2\, dt = -r_I I dt$$
$$\Rightarrow r_I = \frac{r_1 S_1 + r_2 S_2}{S_1 + S_2},$$

and the second moments to be approximately equal,

$$S_1^2\sigma_1^2 + S_2^2\sigma_2^2 + 2\rho S_1\sigma_1 S_2\sigma_2 \cong I^2\sigma_I^2$$
$$\Rightarrow \sigma_I^2 = \frac{S_1^2\sigma_1^2 + S_2^2\sigma_2^2 + 2\rho S_1\sigma_1 S_2\sigma_2}{(S_1 + S_2)^2}.$$

The Black-Scholes equation can therefore be used to value an index option with the above inputs for index borrow and volatility. It gives good qualitative answers for theoretical values and deltas of the option. This is due to the main value of an option being driven by the first and second moments. Note that delta will have extra terms due to the volatility and borrow (vega and ρ_1) dependence:

$$call = C(I, K; r_\$, r_I, \sigma_I, t, T)$$

$$\Delta_1 = \frac{\partial \tilde{C}(S_1, S_2)}{\partial S_1} = \frac{\partial C}{\partial I} + \frac{\partial C}{\partial \sigma_I^2} \frac{\partial \sigma_I^2}{\partial S_1} + \frac{\partial C}{\partial r_I} \frac{\partial r_I}{\partial S_1}$$

The delta for the other stock is a similar expression (take derivative with respect to 2 instead of 1).

Furthermore, for an index with n components,

$$I = \sum_{i=1}^{n} S_i,$$

there are $n(n-1)/2$ cross-correlations ρ_{ij}. We have best-fit index borrow and volatility inputs for the Black-Scholes equation given by

$$r_I = \frac{\sum_{i=1}^{n} r_i S_i}{I},$$

$$\sigma_I^2 = \frac{\sum_{i=1}^{n} S_i^2 \sigma_i^2 + 2 \sum_{i=1}^{n-1} \sum_{j=i+1}^{n} \rho_{ij} S_i \sigma_i S_j \sigma_j}{I^2}.$$

Now, we may conjecture that there are typical or average values for the pairwise correlations and volatilities of the underlying stocks: $\bar{\sigma}, \bar{\rho}$. It follows that

$$\sigma_I^2 = \frac{\bar{\sigma}^2 \left(\sum_{i=1}^{n} S_i^2 + 2\bar{\rho} \sum_{i=1}^{n-1} \sum_{j=i+1}^{n} S_i S_j \right)}{I^2}.$$

Now reinterpret this equation a little more generally: if we assume n equal weight drivers to the index (either stocks or subindexes) of correlation $\bar{\rho}$ to each other and the same volatilities $\bar{\sigma}$, and all the components of each of these subindexes have correlations to each other of 1, then

$$\sigma_I^2 = \frac{\bar{\sigma}^2 \left(\sum_{i=1}^{n} \frac{I^2}{n^2} + 2\bar{\rho} \sum_{i=1}^{n-1} \sum_{j=i+1}^{n} \frac{I^2}{n^2} \right)}{I^2} = \bar{\sigma}^2 \left(\frac{1}{n} + \bar{\rho} \left(1 - \frac{1}{n} \right) \right).$$

Thus for large n, and assuming that all the stocks in the index are the independent drivers and that all have an average pairwise correlation $\overline{\rho}$, we see that the ratio of the index implied volatility to the average component volatility is the average pairwise correlation of all the index's components:

$$\overline{\rho} \approx \frac{\sigma_I^2}{\overline{\sigma}^2}.$$

Alternatively, if we suppose that all stocks are a linear sum of a *small number* of independent drivers (or at least of correlations much lower than the average pairwise correlation of all the index components) and the drivers have similar representation in the index as a whole, then the number of these drivers is

$$n \approx \frac{\overline{\sigma}^2}{\sigma_I^2} \approx \frac{1}{\overline{\rho}}.$$

If typical S&P100 one-month, stock-option-implied volatilities are 35 percent, while typical S&P100 one-month, index-option-implied volatilities are 15 percent, then we may infer that the average implied pairwise correlation of all the index components is $\overline{\rho} \approx 0.36$. This also corresponds to an indication that there are five or six fundamental drivers implied by the dispersion between S&P500 option prices and individual option prices.

If the trader takes a view on correlation (only), then the trade is to buy *at-the-money* (ATM) index options and sell short ATM individual options or *vice versa*. The trade requires making sure that all the component vega (sensitivity to volatility) numbers (for each underlying stock) are zero, and then hedging all the net stock deltas by buying or selling each underlying stock. The result would be (theoretically at least) a pure long, or pure short, correlation position. This is known in the markets as *dispersion trading*.

11.4. MAX OF n STOCKS: TRADING EQUITY CORRELATION

Consider a security that is an option to own either of two stocks. Assume that the two stocks' price changes are lognormally distributed and also correlated as

$$dS_1 = S_1 \mu_1 \, dt + S_1 \sigma_1 dx_1,$$

$$dS_2 = S_2 \mu_2 \, dt + S_2 \sigma_2 dx_2,$$

$$\langle dS_1 dS_2 \rangle = S_1 S_2 \rho_{12} \sigma_1 \sigma_2 \, dt.$$

Then

$$m = \frac{S_1 e^{(-r_1(T-t))}}{S_2 e^{(-r_2(T-t))}},$$

$$dm = m\sigma_2^2 \, dt + m\sigma_m \, dz(t),$$

$$\sigma_m^2 = \sigma_1^2 + \sigma_2^2 - 2\rho\sigma_1\sigma_2.$$

The security terminal value is

$$C_T(m) = S_2 \max(m, 1) = S_2 \max(m - 1, 0) + S_2.$$

Writing the security price in the functional form as

$$C(m, t) = S_2 e^{-r_2(T-t)} F(m, t),$$

we find that while $C(S_1, S_2)$ generally satisfies

$$\frac{\partial C}{\partial t} + \frac{S_1^2 \sigma_1^2}{2} \frac{\partial^2 C}{\partial S_1^2} + \frac{S_2^2 \sigma_2^2}{2} \frac{\partial^2 C}{\partial S_2^2} + S_1 \sigma_1 S_2 \sigma_2 \rho_{12} \frac{\partial^2 C}{\partial S_1 \partial S_2}$$

$$+ S_1(r_\$ - r_1) \frac{\partial C}{\partial S_1} + S_2(r_\$ - r_2) \frac{\partial C}{\partial S_2} - r_\$ C = 0,$$

$F(m, t)$ satisfies

$$\frac{\partial F}{\partial t} + \frac{m^2 \sigma_m^2}{2} \frac{\partial^2 F}{\partial m^2} = 0,$$

and we see that m is a martingale for the process over which F is defined (although not C).

We can then solve the European option (with no-underlying-dividend) case using the Black-Scholes formula:

$$C_{max\ of\ two\ stocks}(S_1, S_2, t)$$

$$= S_2 e^{-r_2(T-t)} [1 + C_{Black-Scholes}(m, 1, t, T, \sigma_m, r_1 = 0, r_2 = 0)];$$

$$m = \frac{S_1 e^{(-r_1(T-t))}}{S_2 e^{(-r_2(T-t))}}, \quad \sigma_m^2 = \sigma_1^2 + \sigma_2^2 - \rho\sigma_1\sigma_2.$$

It clearly has the correct limits for $S_1 \gg S_2$ and $S_1 \ll S_2$. This is an exact result, not an approximation; it assumes the usual Black-Scholes paradigm of lognormally distributed stock price changes together with a constant correlation of changes ρ.

This option versus individual equity options gives a method of trading correlation if the positions of the two individual equity options and the individual equities are chosen to net the vegas and deltas to zero.

As a corollary, let's consider approximating the value of a call on the max of two stocks to which the numerical solution was outlined earlier (section 9.3.6 2-D Models, Correlation and Variable Changes) as the Black-Scholes equation and then using that function involving the Black-Scholes equation itself as the argument. This approximation assumes that the max of two stocks is lognormally distributed, which it is not if the two individual stocks *are* lognormally distributed. But if we fix the first and second moments, borrow and volatility, as

$$dC_{max} = r_\$ C_{max}\, dt + \frac{\partial C_{max}}{\partial S_1} S_1 \sigma_1 dz_1 + \frac{\partial C_{max}}{\partial S_2} S_2 \sigma_2 dz_2 \Rightarrow r_{max} = 0$$

$$\sigma_{max}^2 = \left(\frac{\partial C_{max}}{\partial S_1}\frac{S_1\sigma_1}{C_{max}}\right)^2 + \left(\frac{\partial C_{max}}{\partial S_2}\frac{S_2\sigma_2}{C_{max}}\right)^2$$
$$+ 2\rho\left(\frac{\partial C_{max}}{\partial S_1}\frac{S_1\sigma_1}{C_{max}}\right)\left(\frac{\partial C_{max}}{\partial S_2}\frac{S_2\sigma_2}{C_{max}}\right)$$

then the answer will be very close and clearly works exactly in the limit of either stock getting large:

$$C_{call\ on\ max\ of\ 2} = C_{Black-Scholes}(C_{max}, K, t, T; \sigma_{max}, r_\$, 0)$$

$$\sigma_{max}\ as\ above;$$

and

$$C_{max}(S_1, S_2, t) = S_2 e^{-r_2(T-t)}[1 + C_{Black-Scholes}(m, 1, t, T, \sigma_m, 0, 0)].$$

$$m = \frac{S_1 e^{(-r_1(T-t))}}{S_2 e^{(-r_2(T-t))}}; \quad \sigma_m^2 = \sigma_1^2 + \sigma_2^2 - \rho\sigma_1\sigma_2$$

Quite complex derivatives can often be analytically approximated this way.

11.5. COLLATERALIZED DEBT OBLIGATIONS (CDOs): TRADING CREDIT CORRELATION

There is already quite an extensive literature on *collateralized debt obligation* (CDO) pricing, and the Gaussian Copula model (originally developed by

Li 2000) has become a benchmark. However, without delving into this, this section provides the basic tools of default correlation that lead to the simplest model—taking only bond prices and a complete default rate correlation matrix for inputs. While this "toy" model does not have the same tranche structures as any actual CDO in the market place, the tools used here can be applied to the wide majority of the structures out there. For example, CDO tranches above the equity tranche usually have a prospectus imposed coupon instead of the assumption made here of a proportion of the actual coupon and synthetic CDOs are actually credit default swaps on a CDO tranche backed by a pool of CDS. Both these structures can be modeled using the tools developed here.

We must note the general high degree of difficulty in implementing CDO models numerically: it is the driver of this active field of research. For N bonds we know there are 2^N separate outcomes: survival or default of each bond; and this is a very big number for a 100-bond CDO! On the other hand, simplifying the calculation results in models that are hard to realistically relate (i.e., calibrate) to the real world. So the modeler is left with the classic tradeoff between fast valuation and realistic models.

Commensurately, directly implementing the framework described below is hard numerically but it is much more realistic than the Copula model because it takes into account all bonds pairwise default correlations. At the other end of the spectrum, the Copula model is easier to implement because it assumes that all bonds are not directly correlated to each other but are correlated to a *fictitious* market-wide firm value and thus requires this correlation *vector* and the strikes, or expected times to default, of each bond. These inputs are already the Achilles' heel of the Copula model.

The method of development in this section has been chosen because it makes far fewer assumptions and describes the correct framework in which to understand all models.

CDOs describe a structure with the following general characteristics: a pool of N bonds, for example 10 to 100, held by the issuer. Then the issuer sells tranches that have the following typical characteristics: the highest debt-quality tranche pays coupons and par equal to 10 percent of the pool's total par and coupons and is secured by the last 10 percent of bonds in the pool to default. The next highest quality debt is a tranche that might pay the same coupons and par but is secured by the penultimate 10 percent to default. The tranches are all similarly structured, down to the so-called equity tranche (also called *toxic waste*), which has the same coupons and par but is only secured by the first 10 percent to default.

Besides the usual "sum of the parts is more than the whole" reason for the issuer to manufacture this kind of structure, there are good reasons why a genuine demand might be there for it. A CDO, with only lower quality credit in the pool, will have very high quality upper tranches. This might enable bond funds that are restricted to investment-grade debt only to own debt secured by companies whose credit quality is below that grade.

11.5.1. CDO Backed by Three Bonds

As a warm-up exercise, we will value the tranches of a CDO on three bonds. (It turns out that the generalization to N bonds is easy.) Assuming we have default probability and recovery curves with which to price all three bonds and each bond has a par of 1 and cash flows c^1, c^2, and c^3, then the price of bond 1 is given by

$$B_{T_1}^1(t) = s_{T_1}^1 d_{T_1}^1 + \sum_{j \text{ coupons}} s_{t_j}^1 c_j^1 + R_{T_1}^1,$$

$$d_{T_1}^1 = par \, \exp\left(-\int_{s=t}^{s=T_1} r(s) \, ds\right),$$

$$c_j^1 = c^1 \exp\left(-\int_{s=t}^{s=t_j} r(s) \, ds\right),$$

$$s_{T_1}^1 = \exp\left(-\int_{s=t}^{s=T_1} \frac{\phi^1(s)}{1 - R^1(s)} \, ds\right) = \exp\left(-\int_{s=t}^{s=T_1} p^1(s) \, ds\right),$$

$$R_{T_1}^1 = \int_{u=t}^{u=T_1} p(u) R^1(u) \exp\left(-\int_{s=t}^{s=u} p^1(s) \, ds\right) du.$$

Here $r(t)$ is the riskless interest rate curve, $\phi^1(t)$ is the continuous CDS payment curve on bond 1, $R^1(t)$ is the recovery curve on bond 1, the instantaneous default probability rate is $p(t)$, and the integrated survival probability curve is $s_{T_1}^1(t)$.

Now we can count the various outcomes by the following expansion:

$$(p_t^1 + s_t^1)(p_t^2 + s_t^2)(p_t^3 + s_t^3)$$
$$= p_t^1 p_t^2 p_t^3 + p_t^1 p_t^2 s_t^3 + p_t^1 s_t^2 p_t^3 + p_t^1 s_t^2 s_t^3 + s_t^1 p_t^2 p_t^3 + s_t^1 p_t^2 s_t^3$$
$$+ s_t^1 s_t^2 p_t^3 + s_t^1 s_t^2 s_t^3.$$

The (*expectation value of the*) first term is the probability that all three bonds default. The second term represents the probability that bonds 1 and

2 default and bond 3 survives to maturity. We may group these into the various tranches and their payouts of (schematically) par, that is, 1 or recovery as follows:

$$Tr_1 = (s_t^1 s_t^2 s_t^3 + p_t^1 s_t^2 s_t^3 + s_t^1 p_t^2 s_t^3 + s_t^1 s_t^2 p_t^3 + p_t^1 p_t^2 s_t^3 + p_t^1 s_t^2 p_t^3 + s_t^1 p_t^2 p_t^3)par$$
$$+ (p_t^1 p_t^2 p_t^3)R,$$

$$Tr_2 = (s_t^1 s_t^2 s_t^3 + p_t^1 s_t^2 s_t^3 + s_t^1 p_t^2 s_t^3 + s_t^1 s_t^2 p_t^3)par$$
$$+ (p_t^1 p_t^2 s_t^3 + p_t^1 s_t^2 p_t^3 + s_t^1 p_t^2 p_t^3 + p_t^1 p_t^2 p_t^3)R,$$

$$Tr_3 = (s_t^1 s_t^2 s_t^3)par$$
$$+ (p_t^1 s_t^2 s_t^3 + s_t^1 p_t^2 s_t^3 + s_t^1 s_t^2 p_t^3 + p_t^1 p_t^2 s_t^3 + p_t^1 s_t^2 p_t^3 + s_t^1 p_t^2 p_t^3 + p_t^1 p_t^2 p_t^3)R.$$

The first term in tranche 1 represents the probability that no bonds default; the second, third, and fourth that one bond defaults; the fifth, sixth, and seventh that two default and the last that all three default. We write these eight terms for each tranche j as

$$Tr_j = \sum_{i=1}^{8} Tr_j^i.$$

Now we need to add up the values of each term to get the tranche values. To get the right sum, we can look at the term and assume it is an evenly weighted sum of the payout and coupons, or recovery value of the bonds that survive or default, as follows: take the first term in the first tranche:

$$Tr_1^1 = \tfrac{1}{3}\langle s_t^1 s_t^2 s_t^3 (B_1 + B_2 + B_3)\rangle$$
$$= \tfrac{1}{3}\langle s_t^1 s_t^2 s_t^3 B_1\rangle + \tfrac{1}{3}\langle s_t^1 s_t^2 s_t^3 B_2\rangle + \tfrac{1}{3}\langle s_t^1 s_t^2 s_t^3 B_3\rangle$$

We need to be careful about the meaning of the terms. Take the very first term $\langle s_t^1 s_t^2 s_t^3 B_1\rangle$, which means the probability that all three bonds survive. If there is no correlation of default probability then this is simply the product of the survival probabilities, that is,

$$\langle s_t^1 s_t^2 s_t^3 B_1\rangle = \langle s_t^1\rangle \langle s_t^2\rangle \langle s_t^3 B_1\rangle = \left(s_{T_1}^1 s_{T_1}^2 s_{T_1}^3 d_{T_1}^1 + \sum_{j\,coupons} s_{t_j}^1 s_{t_j}^2 s_{t_j}^3 c_j^1 \right)$$

because they are unrelated (uncorrelated probabilities just multiply together to get the probability of a coincidence). Note that the survival probabilities are integrated to the cash flow dates of the bond being conditionally valued, while the default probability curves (under the integral inside the survival probabilities) are all *understood* to be zero after their respective bond maturities because default of a particular bond cannot occur after its own maturity.

The problem with this simple picture is that default correlation is not accounted for. Default correlation is the central issue for trading CDOs—just as equity correlation is for trading options on stock indices. Consider this integral,

$$\langle s_T^1(t) s_T^2(t) \rangle = \left\langle \exp\left(-\int_{s=t}^{s=T} p^1(s)\, ds\right) \exp\left(-\int_{s=t}^{s=T} p^2(s)\, ds\right)\right\rangle$$

where the instantaneous default probabilities are $p^1(t)\Delta t$ and $p^2(t)\Delta t$. Now we want to introduce a default correlation, and a good place to start is instantaneous default correlation. At every instant, and for each bond, we have a two-state system: default and survival. The distribution has the two probabilities, expressed as

$$\langle survival(i)\rangle = s^i(t)\Delta t, \quad \langle default(i)\rangle = p^i(t)\Delta t = 1 - s^i(t)\Delta t.$$

Next consider the survival probability of two different bonds that have a correlation of 1 and the same default probability. We may write the formulae as

$$\langle survival(i)survival(j)\rangle|_{i=j} = s^i(t)\Delta t,$$

$$\langle survival(i)^2\rangle = s^i(t)\Delta t,$$

$$\langle default(i)^2\rangle = p^i(t)\Delta t$$

because the "square" in this expectation value here implies the calculation of the probability of two bonds defaulting, under the condition that they are perfectly correlated and have the same probability of default. The result is just the probability of one defaulting, hence the formulae. Further it follows that

$$\sigma^i = stdev(survival(i)) = \sqrt{s^i(t)\Delta t(1 - s^i(t)\Delta t)} = \Delta t\sqrt{s^i(t)p^i(t)};$$

$$stdev(default(i)) = \sqrt{p^i(t)\Delta t(1 - p^i(t)\Delta t)} = \Delta t\sqrt{s^i(t)p^i(t)} = \sigma^i.$$

The (pairwise) correlation is defined as

$$\rho_{ij} = \frac{\langle default(i \text{ and } j) \rangle - \langle default(i) \rangle \langle default(j) \rangle}{\sigma^i \sigma^j},$$

which is more understandable once rearranged as

$$\langle default(i \text{ and } j) \rangle = \langle default(i) \rangle \langle default(j) \rangle + \rho \sigma^i \sigma^j.$$

This formula has a very simple interpretation: the probability of simultaneous default of bonds i and j might not be the simple product of their individual probabilities of default due to some kind of interaction: correlation. If one company defaults, it might increase the probability of default of the other company due to mutual dependence (correlation > 0) or it might reduce the probability of default of the other company due to reduced competition (correlation < 0). For perfectly uncorrelated default ($\rho = 0$) the double default probability is just the product of the independent default probabilities. If they are perfectly correlated, then the default of both bonds is just the default probability of one of them, and the formula easily reduces to $\rho = 1$. The perfectly anticorrelated case means that the survival probability of one is the default probability of the other and the probability of both defaulting simultaneously is zero. Again the formula easily reduces to $\rho = -1$ (as long as we consider a finite time-step Δt).

Returning to the required expectation value and expanding the survival probabilities at each instant:

$$\langle s_T^1 s_T^2 \rangle = \left\langle \exp\left(-\int_{s=0}^{s=T} p^1(s)\,ds \right) \exp\left(-\int_{s=0}^{s=T} p^2(s)\,ds \right) \right\rangle$$

$$\approx \prod_{t_i=0}^{t_i=T} \langle (1 - p^1(t_i)\Delta t)(1 - p^2(t_i)\Delta t) \rangle$$

$$\approx \prod_{t_i=0}^{t_i=T} (1 - p^1(t_i)\Delta t - p^2(t_i)\Delta t + \langle p^1(t_i)\Delta t p^2(t_i)\Delta t \rangle).$$

We may then evaluate the simultaneous default probability as

$$\langle p^1(t_i)\Delta t p^2(t_i)\Delta t \rangle = p^1(t_i)\Delta t p^2(t_i)\Delta t$$

$$+ \rho\sqrt{(p^1(t)\Delta t(1 - p^1(t)\Delta t))(p^2(t)\Delta t(1 - p^2(t)\Delta t))}$$

$$\approx \rho\sqrt{p^1(t)p^2(t)}\Delta t + O(\Delta t^2).$$

This leads to the conclusion that there is another term of order Δt to be added to factors in the product of survival probabilities, expressed as

$$\prod_i (1 - p^1(t_i)\Delta t - p^2(t_i)\Delta t + \langle p^1(t_i)\Delta t p^2(t_i)\Delta t\rangle)$$

$$\approx \prod_i (1 - [p^1(t_i) - p^2(t_i) + \rho^{12}(t)\sqrt{p^1(t_i)p^2(t_i)}]\Delta t).$$

Finally,

$$\langle s_T^1(t)s_T^2(t)\rangle = \left\langle \exp\left(-\int_{s=t}^{s=T} p^1(s)\,ds\right)\exp\left(-\int_{s=t}^{s=T} p^2(s)\,ds\right)\right\rangle$$

$$= \exp\left(-\int_{s=t}^{s=T} [p^1(s) + p^2(s) - \rho^{12}(s)\sqrt{p^1(s)p^2(s)}]\,ds\right).$$

Note that the correlation of the integrated survival probabilities is

$$\rho_T^{12}(t) = \frac{\langle s_T^1(t)s_T^2(t)\rangle - \langle s_T^1(t)\rangle\langle s_T^2(t)\rangle}{\Sigma_T^1(t)\Sigma_T^2(t)}$$

$$= \frac{\left[\exp\left(\int_{s=t}^{s=T}\rho^{12}(s)\sqrt{p^1(s)p^2(s)}\,ds\right) - 1\right]}{\sqrt{\left[\exp\left(\int_{s=t}^{s=T} p^1(s)\,ds\right) - 1\right]\left[\exp\left(\int_{s=t}^{s=T} p^2(s)\,ds\right) - 1\right]}}.$$

where the integrated variances are easily seen to be the expressions in the numerator for $\rho^{12}(t) = 1$ and $p^1(t) = p^2(t)$. This formula for the correlation of the integrated survival probabilities has the following properties that are default probability rate independent:

$$\lim_{T\to t}\rho_T^{12}(t) = \rho^{12}(t),$$

$$\rho_T^{12}(t)\big|_{\substack{\rho^{12}(t)=1,\\ p^1(t)=p^2(t)}} = 1,$$

$$\rho_T^{12}(t)\big|_{\rho^{12}(t)=0} = 0.$$

We can generalize to many integrated survival probabilities that

$$\langle s_t^1 s_t^2 s_t^3 \ldots s_t^K\rangle^N = \exp\left(-\int_{s=0}^{s=t}\left[\sum_{i=1}^K p^i(s) - C_K^N(\{\rho_{ij}\}; \{p_i\})\right]\,ds\right).$$

See Appendix D for the full expression for $C_K^N(\{\rho_{ij}\}; \{p_i\})$, which is generally a function of all the bonds and correlations in the system—although, for example, $C_2^N(\{\rho_{ij}\}; \{p_i\}) = \rho_{12}\sqrt{p_1 p_2}$ for any two bonds out of an N bond system.

Now the first term may be evaluated easily as follows:

$$\langle s_t^1 s_t^2 s_t^3 B_1 \rangle = \left(\langle s_{T_1}^1 s_{T_1}^2 s_{T_1}^3 \rangle d_{T_1}^1 + \sum_{j \text{ coupons}} \langle s_{t_j}^1 s_{t_j}^2 s_{t_j}^3 \rangle c_j^1 \right)$$

simply using the formulae shown thus far in this section.

The second term of tranche 1 is

$$Tr_1^2 \sim \langle p_t^1 s_t^2 s_t^3 par \rangle$$

schematically where the par payout is again an average over the surviving bonds payouts to get

$$Tr_1^2 = \tfrac{1}{2}\langle p_t^1 s_t^2 s_t^3 (B_2 + B_3) \rangle = \tfrac{1}{2}\langle p_t^1 s_t^2 s_t^3 B_2 \rangle + \tfrac{1}{2}\langle p_t^1 s_t^2 s_t^3 B_3 \rangle.$$

The (integrated) default probability can be replaced by 1 minus the integrated survival probability,

$$\langle p_t^1 s_t^2 s_t^3 B_2 \rangle = \langle s_t^2 s_t^3 B_2 \rangle - \langle s_t^1 s_t^2 s_t^3 B_2 \rangle.$$

Both these terms have been evaluated here already. We can continue to value all par payouts using these rules.

Finally the eighth term in the first tranche values the contribution from recovery, expressed as

$$Tr_1^8 \sim \langle p_t^1 p_t^2 p_t^3 \rangle R,$$

which again we can average over the recovery value of the bonds in default,

$$Tr_1^8 = \tfrac{1}{3}\langle p_t^1 p_t^2 p_t^3 (B_1 + B_2 + B_3) \rangle.$$

Looking at the first term,

$$Tr_1^8 = \tfrac{1}{3}\langle p_t^1 p_t^2 p_t^3 B_1 \rangle,$$

we interpret this term to mean bond 1 survives to date u and then bond 1 defaults while during this time both bond 2 and 3 default. We add up all these contributions for all $u : t < u < T$. So there are many contributions to this term—from default at each instant between today and maturity of bond 1—and it is easier to manipulate the nondefaulting bonds into survival probabilities,

$$
\begin{aligned}
Tr_1^8 &= \tfrac{1}{3}\langle p_t^1(1 - s_t^2)(1 - s_t^3)B_1\rangle \\
&= \tfrac{1}{3}\langle p_t^1 B_1\rangle + \tfrac{1}{3}\langle p_t^1 s_t^2 s_t^3 B_1\rangle - \tfrac{1}{3}\langle p_t^1 s_t^2 B_1\rangle - \tfrac{1}{3}\langle p_t^1 s_t^3 B_1\rangle.
\end{aligned}
$$

The first term we know already as

$$
\langle p_t^1 B_1\rangle = \int_{u=t}^{u=T} p(u)R^1(u)\exp\left(-\int_{s=t}^{s=u} r(s) + p^1(s)\, ds\right) du.
$$

Further note that

$$
\frac{d}{dt}\langle s_t^1 s_t^2 s_t^3 \ldots s_t^K\rangle^N = -\left[\sum_{i=1}^{K} p^i(t) - C_K^N(\{\rho_{ij}(t)\}; \{p_i(t)\})\right]\langle s_t^1 s_t^2 s_t^3 \ldots s_t^K\rangle^N
$$

differentiating the expression for the conditional survival probability above. The rate of change of this survival probability has contributions from many terms; one default, two defaults and so on for all permutations to N defaults at time t. This rate, the probability of not all bonds surviving interval dt at time t, is multiplied by the probability of all surviving to time t. Ignoring bond K we find the probability of not all bonds surviving interval dt from the set of bonds excluding bond K,

$$
\begin{aligned}
\frac{d}{dt}\langle s_t^1 s_t^2 s_t^3 \ldots s_t^{K-1}\rangle^N &= -\left[\sum_{i=1}^{K-1} p^i(t) - C_{K-1}^N(\{\rho_{ij}(t)\}; \{p_i(t)\})\right] \\
&\quad \langle s_t^1 s_t^2 s_t^3 \ldots s_t^{K-1}\rangle^N
\end{aligned}
$$

and so the probability of bond K not surviving interval dt is

$$
\left[\sum_{i=1}^{K} p^i(t) - C_K^N(\{\rho_{ij}(t)\}; \{p_i(t)\})\right] - \left[\sum_{i=1}^{K-1} p^i(t) - C_{K-1}^N(\{\rho_{ij}(t)\}; \{p_i(t)\})\right].
$$

The conclusion is

$$\langle p_t^1 s_t^2 s_t^3 B_1 \rangle$$

$$= \int_{u=t}^{u=T} (F^{123}(u) - F^{23}(u)) R^1(u) \exp\left(-\int_{s=t}^{s=u} [r(s) + F^{123}(s)] \, ds\right) \, du,$$

$$F^{123}(t) = p^1(t) + p^2(t) + p^3(t) - C_3^3(\{p^1(t), p^2(t), p^3(t)\}; \{\rho^{ij}(s)\}),$$

$$F^{23}(t) = p^2(t) + p^3(t) - C_2^3(\{p^2(t), p^3(t)\}; \{\rho^{ij}(s)\})$$

represents the probability that all bonds survive to date u and then only bond 1 defaults. (The terms C_3^3 and C_2^3 are detailed in Appendix D.) Generally, a term of this form, the coincident survival probability of K correlated bonds, in an N correlated bond system, is given by

$$\langle s_t^1 s_t^2 s_t^3 \dots s_t^{K-1} p_t^K B_K \rangle^N$$

$$= \int_{u=t}^{u=T} (F_K^N(u) - F_{K-1}^N(u)) R^K(u) \exp\left(-\int_{s=t}^{s=u} [r(s) + F_K^N(u)] \, ds\right) \, du,$$

$$F_K^N(u) = \sum_{i=1}^{K} p^i(s) - C_K^N(\{\rho^{ij}(s)\}; \{p^i(s)\}),$$

$$F_{K-1}^N(u) = \sum_{i=1}^{K-1} p^i(s) - C_{K-1}^N(\{\rho^{ij}(s)\}; \{p^i(s)\}).$$

This solves the problem of valuing each of the three possible tranches of a CDO backed by a pool of three bonds.

11.5.2. CDO Backed by an Arbitrary Number of Bonds

For N bonds in the pool, we value the N tranches of a CDO. The bonds each have a par of 1 and cash flows c^k, for $1 \leq k \leq N$, which imply prices and default and recovery curves, expressed as

$$B_{T_k}^k(t) = s_{T_k}^k d_{T_k}^k + \sum_{j \text{ coupons}} s_{t_j}^k c_j^k + R_{T_k}^k,$$

$$d_{T_k}^k = par^k \exp\left(-\int_{s=t}^{s=T_k} r(s) \, ds\right),$$

$$c_j^k = c^k \exp\left(-\int_{s=t}^{s=t_j} r(s)\,ds\right),$$

$$s_{T_1}^k = \exp\left(-\int_{s=t}^{s=T_k} \frac{\phi^k(s)}{1 - R^k(s)}\,ds\right) = \exp\left(-\int_{s=t}^{s=T_k} p^k(s)\,ds\right),$$

$$R_{T_1}^k = \int_{u=t}^{u=T_k} p^k(u) R^k(u) \exp\left(-\int_{s=t}^{s=u} [r(s) + p^k(s)]\,ds\right)\,du.$$

Here $r(t)$ is the riskless interest rate curve, $\phi^k(t)$ is the continuous CDS payment curve on bond k, $R^k(t)$ is the recovery curve on bond k, the instantaneous default probability rate is $p^k(t)$, and the integrated survival probability of bond k is $s_{T_1}^k(t)$.

The various outcomes; combinations of default or survival of each bond, may be counted using the expansion

$$
\begin{aligned}
(p_t^1 + s_t^1)(p_t^2 + s_t^2)(p_t^3 + s_t^3)\ldots(p_t^N + s_t^N) &= s_t^1 s_t^2 s_t^3 \ldots s_t^N \\
&+ p_t^1 s_t^2 s_t^3 \ldots s_t^N + s_t^1 p_t^2 s_t^3 \ldots s_t^N + \cdots + s_t^1 s_t^2 s_t^3 \ldots p_t^N \\
&+ p_t^1 p_t^2 s_t^3 \ldots s_t^N + p_t^1 s_t^2 p_t^3 \ldots s_t^N + \cdots + s_t^1 s_t^2 s_t^3 \ldots s_t^{N-2} p_t^{N-1} p_t^N \\
&\vdots \\
&+ p_t^1 p_t^2 p_t^3 \ldots p_t^N
\end{aligned}
$$

and the 2^N terms have been arranged into groups with increasing number of defaults (same number of defaults for each line). Then the N tranches may be cast as follows:

$$
\begin{aligned}
Tr_1 &= (s_t^1 s_t^2 s_t^3 \ldots s_t^N + s_t^1 p_t^2 s_t^3 \ldots s_t^N + \cdots + s_t^1 p_t^2 p_t^3 \ldots p_t^N)par \\
&\quad + (p_t^1 p_t^2 p_t^3 \ldots p_t^N)R, \\
Tr_2 &= (s_t^1 s_t^2 s_t^3 \ldots s_t^N + \cdots + s_t^1 s_t^2 p_t^3 \ldots p_t^N)par \\
&\quad + (s_t^1 p_t^2 p_t^3 \ldots p_t^N + p_t^1 p_t^2 p_t^3 \ldots p_t^N)R, \\
&\vdots \\
Tr_j &= (s_t^1 s_t^2 s_t^3 \ldots s_t^N + \cdots + s_t^1 \ldots s_t^j p_t^{j+1} \ldots p_t^N)par \\
&\quad + (s_t^1 \ldots s_t^{j-1} p_t^j \ldots p_t^N + p_t^1 p_t^2 p_t^3 \ldots p_t^N)R, \\
&\vdots \\
Tr_N &= (s_t^1 s_t^2 s_t^3)par + (p_t^1 s_t^2 s_t^3 \ldots s_t^N + \cdots + p_t^1 p_t^2 p_t^3 \ldots p_t^N)R.
\end{aligned}
$$

The first group of terms in the jth tranche represents the probability of payout of the portion of par that the holder of this tranche has a claim to. This happens when there are j defaults or less of bonds in the pool. The second term values the holder's claim in default (recovery). This payout happens when j or more bonds default. In order to value each term, we will write each tranche as

$$Tr_j = \sum_{i=1}^{i=2^N} Tr_j^i,$$

where each term Tr_j^i is to be read off the above expansion.

To value these terms we treat the tranche *par* terms and *recovery* terms slightly differently. The par terms need to be the expectation value of an average of the surviving bonds,

$$Tr_j^i = \langle s_t^1 \ldots s_t^k p_t^{k+1} \ldots p_t^N par \rangle = \frac{1}{m} \langle s_t^1 \ldots s_t^k p_t^{k+1} \ldots p_t^N (B_1 + \cdots B_m) \rangle$$

and this generates terms of the form

$$\langle s_t^1 \ldots s_t^k p_t^{k+1} \ldots p_t^N B_j \rangle |_{1 \leq j \leq k} - \langle s_t^1 \ldots s_t^k (1 - s_t^{k+1}) \ldots (1 - s_t^N) B_j \rangle |_{1 \leq j \leq k},$$

which in turn generates terms of the form

$$\langle s_t^1 \ldots s_t^k s_t^{k+1} \ldots s_t^m B_j \rangle |_{1 \leq j \leq k} = \left(\langle s_{T_j}^1 \ldots s_{T_j}^m \rangle par^j + \sum_{i\ coupons} \langle s_{t_i}^1 \ldots s_{t_i}^m \rangle c^i \right)$$

for $m \leq N$. These terms are valued using the expression

$$\langle s_t^1 s_t^2 s_t^3 \ldots s_t^K \rangle^N = \exp \left(- \int_{s=0}^{s=t} \left[r(s) + \sum_{i=1}^K p^i(s) - C_K^N(\{\rho_{ij}\}; \{p_i\}) \right] ds \right)$$

where the riskless discounting term is now incorporated into the expectation value of the survival probability, so that the par^j and c^i values are just the cash flows of the jth bond. See Appendix D for the full expression including the method of calculating $C_K^N(\{\rho_{ij}\}; \{p_i\})$.

The *recovery* terms in the tranches are of the form

$$\langle p_t^1 \ldots p_t^k s_t^{k+1} \ldots s_t^N \rangle rec = \frac{1}{k} \langle p_t^1 \ldots p_t^k s_t^{k+1} \ldots s_t^N (B_1 + B_2 + \cdots + B_k) \rangle |_{1 \leq j \leq k}$$

and each of these terms generates terms of the form

$$\langle p_t^1 \dots p_t^k s_t^{k+1} \dots s_t^N B_j \rangle$$

$$= \langle (1 - s_t^1) \dots (1 - s_t^{j-1}) p_t^{j+1} (1 - s_t^{j+1}) \dots (1 - s_t^k) s_t^{k+1} \dots s_t^N B_j \rangle |_{1 \leq j \leq k}$$

which, in turn, generates terms of the form

$$\langle s_t^1 s_t^2 s_t^3 \dots s_t^{K-1} p_t^K B_K \rangle^N$$

$$= \int_{u=t}^{u=T} (F_K^N(u) - F_{K-1}^N(u)) R^K(u) \exp \left(- \int_{s=t}^{s=u} [r(s) + F_K^N(u)] \, ds \right) \, du,$$

$$F_K^N(u) = \sum_{i=1}^{K} p^i(s) - C_K^N(\{\rho^{ij}(s)\}; \{p^i(s)\}),$$

$$F_{K-1}^N(u) = \sum_{i=1}^{K-1} p^i(s) - C_{K-1}^N(\{\rho^{ij}(s)\}; \{p^i(s)\}).$$

The sum of all these terms with their weights gives the jth subtranche Tr_j of the actual CDO. Adding up these subtranches gives actual CDO tranches. For example, an actual tranche might be backed by the last 10 percent to default on a pool of 100 bonds and its value will be the sum of the first 10 subtranches. The form of the term $C_K^N(\{\rho_{ij}(s)\}; \{p_i(s)\})$ is detailed in Appendix D.

Note last that CDO tranche issues are OTC (over-the-counter) and generally all slightly different with set coupon payments for non-equity tranches and coupon coverage requirements and various bells and whistles that require specific tinkering with a model for each CDO and even each tranche. Furthermore, synthetic CDO tranches are more like credit default swaps. The machinery of the above calculation can be extended to these cases.

The numerical implementation of this type of algorithm (and good CDO tranche models generally) requires optimization and some serious computing power. After all, there are 2^N terms for a CDO with N bonds in the pool. A simple optimization is to compute the terms in order of number of bonds defaulting and track the contribution of the sums of these terms to the tranche value and stop at some tranche evaluation accuracy level. This effectively makes the higher credit quality tranches the same. This highlights that the higher quality tranches are similar high investment grade and, as the tranche level decreases, the credit quality suddenly drops off until the equity tranche is very low-credit quality and near to the recovery value of the entire pool.

PART II

Exercises and Solutions

CHAPTER 12

Exercises

Exercise 1 (Section 2.1)

If

$$p(x) = \frac{1}{\sqrt{2\pi}} \exp\left(-\frac{x^2}{2}\right),$$

what is $\int_{-\infty}^{\infty} p(x)\,dx$? Then what is $\langle x \rangle$, $\langle x^2 \rangle$, and $\langle \exp(x) \rangle$?

Exercise 2 (Section 2.1)

Show that the convolution of two Gaussian distributions,

$$p(x) = \frac{1}{\sqrt{2\pi}} \exp\left(-\frac{x^2}{2}\right)$$

$$[p * p](x) = \int_{y=-\infty}^{y=\infty} p(y-x)p(y)\,dy,$$

is also a Gaussian.

Exercise 3 (Section 2.1)

Given the sets of random variables

$$\langle z_1 \rangle = 0,$$
$$\langle z_2 \rangle = 0,$$
$$\mathrm{var}(z_1) = 1,$$
$$\mathrm{var}(z_2) = 1,$$
$$\langle z_1 z_2 \rangle = 0,$$

we are free to choose α_1, α_2, and β so that

$$x_1 = \alpha_1(z_1 + \beta z_2) + \mu_1,$$

$$x_2 = \alpha_2(z_1 - \beta z_2) + \mu_2.$$

What are

$$\langle x_1 \rangle, \langle x_2 \rangle, \langle x_1^2 \rangle, \langle x_2^2 \rangle, \langle x_1^2 \rangle - \langle x_1 \rangle^2, \langle x_2^2 \rangle - \langle x_2 \rangle^2, \sigma_1, \sigma_2, \langle x_1 x_2 \rangle, \rho?$$

Exercise 4 (Section 2.1)

Write out α_1, α_2, β as functions of μ_1, μ_2, σ_1, σ_2, ρ, and vice versa, μ_1, μ_2, σ_1, σ_2, ρ as functions of α_1, α_2, β.

Exercise 5 (Section 2.1)

Write out the distribution function for $p(x_1, x_2)\, dx_1\, dx_2$. *Hint:* just a change of variables from $p(z_1, z_2)\, dz_1\, dz_2$.

Exercise 6 (Section 2.2)

Does $f(x, y)$ exist for the following? If so, what is it?

1. $df = y\sin(\omega xy)\, dx + y\cos(\omega xy)\, dy.$
2. $df = y\cos(\omega xy)\, dx + x\cos(\omega xy)\, dy.$

Reminder: Because

$$\frac{\partial^2 f}{\partial x \partial y} = \frac{\partial^2 f}{\partial y \partial x},$$

then, for $df = h(x, y)\, dx + g(x, y)\, dy$, it must be true that

$$\frac{\partial h(x, y)}{\partial y} = \frac{\partial g(x, y)}{\partial x}.$$

This is the test of existence of $f(x, y)$. If f exists then solve the following:

$$\frac{\partial f}{\partial x} = h(x, y),$$

$$\frac{\partial f}{\partial y} = g(x, y).$$

Exercise 7 (Section 3.1)

Find the mean and standard deviation of X if it is defined by the stochastic integral

$$X = \int_{s=t}^{s=T} f(s)\,ds + g(s)\,dz(s)$$

where $f(t)$ and $g(t)$ are arbitrary (nonstochastic) functions of time t and $dz(t)$ is a Wiener process.

Exercise 8 (Section 3.2)

Write out the distribution for X, and don't forget the measure (volume element).

Exercise 9 (Section 3.2)

Given a process for x,

$$dx = \mu(x,t)\,dt + \sigma(x,t)\,dz(t),$$

consider the variable $S = e^x$. Derive the process for S using Ito's lemma.

Exercise 10 (Section 4.1)

Given a market with two drivers, $dz_1(t)$ and $dz_2(t)$, investigate the generalization of the risk premium.

Exercise 11 (Section 4.2)

1. What is the probability that the stock ends up above strike for a call option?
2. Where does this probability show up in the call expectation value formula?

Exercise 12 (Section 5.1)

For the process

$$dS = S\mu_0\,dt + S\sigma_0\,dz(t),$$

find functions $f(T, S_T)$ that are solutions to the corresponding forward Kolmogorov equation and that have the initial value S_t^n at time $T = t$. Assume $f(T, S_T)$ is of the form, $f(T)S_T^n$, for n equals 1, 2 and 3.

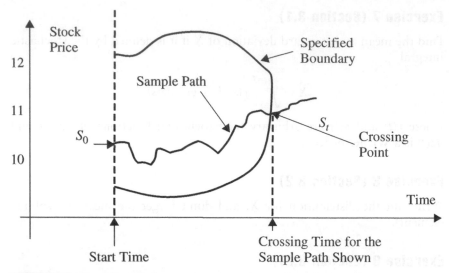

FIGURE 12.1 For all paths beginning at So, what is the average (expectation) time to cross a specified boundary?

Exercise 13 (Section 5.1)

1. Find functions $f(t, S_t)$ that are solutions to the backward Kolmogorov equation and that have the initial value S_T^n at time $t = T$. Assume $f(t, S_t)$ is of the form, $f(t)S_t^n$, for n equals 1, 2, and 3.
2. Relate these six functions to the original process.

Exercise 14 (Section 5.1)

How would one calculate the expected time to cross an arbitrary boundary, having started at S_t as shown in Figure 12.1, that is, $\langle T - t \rangle = f(S_t; [boundary])$?

Exercise 15 (Section 5.1)

How would one calculate the expected time for a path ending at S_T, to arrive from a given stock price boundary, that is, $\langle T - t \rangle = f(S_T; [boundary])$?

Exercise 16 (Section 5.2)

If the risk premium has a stochastic process, what is the effect on the Black-Scholes pricing equation for call options?

Exercise 17 (Section 5.2)

We have the formula for the expectation value of the hedging strategy.

$$\langle \text{Hedging strategy} \rangle = E(payoff, \lambda = 0) - E(payoff, \lambda = 0).$$

How would you calculate the variance of the hedging strategy?

Exercise 18 (Section 4.1, 5.2, 6.2, 6.3)

The result that an option price written as a function of stock price, $C(S,t)$, obeys a pricing equation very similar to that solved by a stock price written as a function of a derivative price, $S(C,t)$, was apparent in section 4.1 Risk Premium Derivation and section 6.3 Black-Scholes Equation: Relation to Risk Premium Definition. Make this argument.

Exercise 19 (Section 9.1)

Switch the following operator into coordinates (variables) so that the second order term has a number not a function multiplying it:

$$\left(\frac{\partial}{\partial t} + \frac{S^2 \sigma^2}{2} \frac{\partial^2}{\partial S^2} + S\mu \frac{\partial}{\partial S} - r_\$ \right) = 0.$$

Exercise 20 (Section 9.2)

What is the Fourier transform of a put payout?

Exercise 21 (Section 9.2)

What is the Fourier transform of a put formula for all times?

Exercise 22 (Section 9.3.5)

If we use the European call option formula to price an American call option by making the maturity a function of stock price and time $T(S,t)$, what is the equation satisfied by this function?

Exercise 23 (Section 10.2)

For flat forward curves, credit curves, and recovery curves evaluate the recovery term for a straight bond.

Exercise 24 (Section 11.5)

Show that the correlated survival probability of K bonds—

$$\langle s_1 s_2 \ldots s_K \rangle^{(N)} = \exp\left(-\int F_K^{(N)}\, dt\right)$$

in a system of N (coincidentally) identical bonds with the same credit curves $\frac{\phi}{1-R}$ and pairwise correlations ρ—is given by

$$F_K^{(N)} = \frac{(N - (N-K)(1-\alpha)^K)}{(1+(N-1)\alpha)} \frac{\phi}{(1-R)},$$

where α is given by

$$\rho = \frac{2\alpha + (N-2)\alpha^2}{1 + (N-1)\alpha}.$$

CHAPTER 13

Solutions

Solution to Exercise 1

$$\int_{-\infty}^{\infty} p(x)\, dx = 1,$$

$$\langle x \rangle = \int_{-\infty}^{\infty} x p(x)\, dx = 0,$$

$$\langle x^2 \rangle = \int_{-\infty}^{\infty} x^2 p(x)\, dx = 1,$$

$$\langle \exp(x) \rangle = \int_{-\infty}^{\infty} \exp(x) p(x)\, dx = \exp\left(\frac{1}{2}\right).$$

Observables will tend toward these numbers on average as the number of observations goes to infinity. If X_i is drawn from the distribution of x, $p(x)$, then

$$\lim_{N \to \infty} \left[\frac{1}{N} \sum_{i=1}^{i=N} X_i \right] = \langle x \rangle = 0,$$

$$\lim_{N \to \infty} \left[\frac{1}{N} \sum_{i=1}^{i=N} X_i^2 \right] = \langle x^2 \rangle = 1,$$

$$\lim_{N \to \infty} \left[\frac{1}{N} \sum_{i=1}^{i=N} \exp(X_i) \right] = \langle \exp(x) \rangle = \exp\left(\frac{1}{2}\right).$$

Solutions to Exercises 2 through 5

$$p(x) = \frac{1}{\sqrt{2\pi}} \exp\left(-\frac{x^2}{2}\right),$$

$$[p * p](x) = \int_{y=-\infty}^{y=\infty} p(y - x)p(y)\, dy$$

$$= \left(\frac{1}{\sqrt{2\pi}}\right)^2 \int_{y=-\infty}^{y=\infty} \exp\left(-\frac{(y-x)^2}{2}\right) \exp\left(-\frac{y^2}{2}\right) dy$$

$$= \left(\frac{1}{\sqrt{2\pi}}\right)^2 \int_{y=-\infty}^{y=\infty} \exp\left(-\frac{(y-x/2)^2}{2}\right) \exp\left(-\frac{(y+x/2)^2}{2}\right) dy$$

$$= \left(\frac{1}{\sqrt{2\pi}}\right)^2 \int_{y=-\infty}^{y=\infty} \exp\left(-y^2 - \frac{x^2}{4}\right) dy$$

$$= \left(\frac{1}{\sqrt{2\pi}}\right)^2 \exp\left(-\frac{x^2}{4}\right) \int_{y=-\infty}^{y=\infty} \exp(-y^2)\, dy$$

$$= \left(\frac{1}{\sqrt{2\pi}}\right)^2 \exp\left(-\frac{x^2}{4}\right) \int_{y=-\infty}^{y=\infty} \exp\left(-\frac{y^2}{2}\right) \frac{dy}{\sqrt{2}}$$

$$= \left(\frac{1}{\sqrt{2\pi}}\right)^2 \exp\left(-\frac{x^2}{4}\right) \frac{\sqrt{2\pi}}{\sqrt{2}}$$

$$= \frac{1}{\sqrt{2\pi(\sqrt{2})^2}} \exp\left(-\frac{x^2}{2(\sqrt{2})^2}\right).$$

The result is a Gaussian of width $\sigma = \sqrt{2}$ in the variable x.
First, based on

$$\langle z_1 \rangle = 0,\ \langle z_2 \rangle = 0,\ \langle z_1^2 \rangle = 1,\ \langle z_2^2 \rangle = 1,\ \langle z_1 z_2 \rangle = 0$$

we can assign expectation values of functions of z_1, z_2 very easily as follows:

$$\langle x_1 \rangle = \mu_1,$$

$$\langle x_2 \rangle = \mu_2,$$

$$\langle (x_1 - \mu_1)^2 \rangle = \alpha_1^2(1 + \beta^2),$$

$$\text{note:}\quad \langle x_1^2 \rangle = \langle (x_1 - \mu_1)^2 \rangle + \mu_1^2,$$

$$\langle (x_2 - \mu_2)^2 \rangle = \alpha_2^2(1 + \beta^2),$$

$$\text{let}\quad \sigma_1 = \alpha_1\sqrt{1 + \beta^2},\quad \sigma_2 = \alpha_2\sqrt{1 + \beta^2},$$

$$\rho = \frac{\langle x_1 x_2 \rangle - \langle x_1 \rangle \langle x_2 \rangle}{\sigma_1 \sigma_2} = \frac{1 - \beta^2}{1 + \beta^2}.$$

We can see that the variables x_1, x_2 *are* correlated, even though the variables z_1, z_2 are not. These new variables (x_1, x_2) have means, standard deviations and correlation of μ_1, μ_2, σ_1, σ_2 and ρ, respectively. Inverting we get the following:

$$\alpha_1 = \sigma_1 \sqrt{\frac{1+\rho}{2}},$$

$$\alpha_2 = \sigma_2 \sqrt{\frac{1+\rho}{2}},$$

$$\beta = \sqrt{\frac{1-\rho}{1+\rho}}.$$

(z_1, z_2) are distributed according to

$$p(z_1, z_2)dz_1 dz_2 = \frac{1}{2\pi} \exp\left(-\frac{z_1^2 + z_2^2}{2}\right) dz_1 dz_2.$$

The variables (x_1, x_2), after a simple change of variables from (z_1, z_2), become

$$z_1 = \frac{1}{2}\left[\left(\frac{x_1 - \mu_1}{\alpha_1}\right) + \left(\frac{x_2 - \mu_2}{\alpha_2}\right)\right],$$

$$z_2 = \frac{1}{2\beta}\left[\left(\frac{x_1 - \mu_1}{\alpha_1}\right) - \left(\frac{x_2 - \mu_2}{\alpha_2}\right)\right],$$

where α_1, α_2 and β were given before and the only subtlety is the volume element

$$dz_1 dz_2 = \frac{1}{2}\left[\left(\frac{dx_1}{\alpha_1}\right) + \left(\frac{dx_2}{\alpha_2}\right)\right] \wedge \frac{1}{2\beta}\left[\left(\frac{dx_1}{\alpha_1}\right) - \left(\frac{dx_2}{\alpha_2}\right)\right]$$

$$= dx_1 dx_2 \frac{1}{4\beta}\left[-\left(\frac{1}{\alpha_1}\right)\left(\frac{1}{\alpha_2}\right) - \left(\frac{1}{\alpha_1}\right)\left(\frac{1}{\alpha_2}\right)\right] = -dx_1 dx_2 \frac{1}{2\alpha_1\alpha_2\beta},$$

which shows that the Jacobian is

$$\frac{1}{2\alpha_1\alpha_2\beta} = \frac{1}{\sigma_1\sigma_2\sqrt{1-\rho^2}}.$$

(We can ignore an overall minus sign because we know probability distributions always have a positive sign.)

Finally

$$p(x_1, x_2)dx_1 dx_2 = \frac{dx_1 dx_2}{2\pi\sigma_1\sigma_2\sqrt{1-\rho^2}} \exp\left[-\frac{1}{2(1-\rho^2)}\left\{\left(\frac{x_1-\mu_1}{\sigma_1}\right)^2\right.\right.$$
$$\left.\left. + \left(\frac{x_2-\mu_2}{\sigma_2}\right)^2 - 2\rho\left(\frac{x_1-\mu_1}{\sigma_1}\right)\left(\frac{x_2-\mu_2}{\sigma_2}\right)\right\}\right].$$

Solution to Exercise 6

1. $f(x,y)$ does not exist.
2. $f(x,y) = \frac{\sin(\omega xy)}{\omega}$.

Solution to Exercise 7

Use the expressions

$$\langle dz(t)\rangle = 0,$$

$$\langle dz(t)\, dz(s)\rangle = dtds\delta(s-t),$$

to evaluate

$$\langle X\rangle = \left\langle \int_{s=t}^{s=T} f(s)\, ds + g(s)\, dz(s) \right\rangle$$

$$= \int_{s=t}^{s=T} \langle f(s)\, ds + g(s)\, dz(s)\rangle$$

$$= \int_{s=t}^{s=T} \langle f(s)\, ds\rangle + \langle g(s)\, dz(s)\rangle$$

$$= \int_{s=t}^{s=T} f(s)\, ds + g(s)\langle dz(s)\rangle$$

$$= \int_{s=t}^{s=T} f(s)\, ds,$$

and

$$\langle X^2\rangle = \left\langle \left[\int_{s=t}^{s=T} f(s)\, ds + g(s)\, dz(s)\right]\left[\int_{u=t}^{u=T} f(u)\, du + g(u)\, dz(u)\right]\right\rangle$$

$$= \left\langle \int_{s=t}^{s=T} f(s)\, ds \int_{u=t}^{u=T} f(u)\, du \right\rangle + \left\langle \int_{s=t}^{s=T} g(s)\, dz(s) \int_{u=t}^{u=T} f(u)\, du \right\rangle$$

$$+ \left\langle \int_{s=t}^{s=T} f(s)\, ds \int_{u=t}^{u=T} g(u)\, dz(u) \right\rangle + \left\langle \int_{s=t}^{s=T} g(s)\, dz(s) \int_{u=t}^{u=T} g(u)\, dz(u) \right\rangle$$

$$= \int_{s=t}^{s=T} f(s)\, ds \int_{u=t}^{u=T} f(u)\, du + \int_{s=t}^{s=T} g(s)\langle dz(s)\rangle \int_{u=t}^{u=T} f(u)\, du$$

$$+ \int_{s=t}^{s=T} f(s)\, ds \int_{u=t}^{u=T} g(u)\langle dz(u)\rangle + \int_{u=t}^{u=T} \int_{s=t}^{s=T} g(s)g(u)\langle dz(s)\, dz(u)\rangle$$

$$= \int_{s=t}^{s=T} f(s)\, ds \int_{u=t}^{u=T} f(u)\, du + \int_{u=t}^{u=T} \int_{s=t}^{s=T} g(s)g(u)\delta(s-u)\, ds\, du$$

$$= \left(\int_{s=t}^{s=T} f(s)\, ds \right)^2 + \int_{s=t}^{s=T} g^2(s)\, ds.$$

It follows that the mean and variance (standard deviation is the square root of the variance) of the stochastic integral, for X, are

$$\text{mean} = \langle X \rangle = \int_{s-t}^{s=T} f(s)\, ds,$$

$$\text{variance} = \langle X^2 \rangle - \langle X \rangle^2 = \int_{s=t}^{s-T} g^2(s)\, ds.$$

Solution to Exercise 8

X is normally distributed with mean and variance given above. We may write a mean 0, variance 1, distribution as

$$p(z)\, dz = \frac{1}{\sqrt{2\pi}} \exp\left(-\frac{z^2}{2} \right) dz,$$

which implies, under the variable change $z = \frac{x-\mu}{\sigma}$, that we get

$$p(x)\, dx = \frac{1}{\sqrt{2\pi\sigma^2}} \exp\left(-\frac{(x-\mu)^2}{2\sigma^2} \right) dx,$$

which is a mean μ, variance σ^2 distribution. Thus the distribution for X may be written

$$p(X)\,dX = \frac{1}{\sqrt{2\pi\sigma^2}}\exp\left(-\frac{(X-\mu)^2}{2\sigma^2}\right)dX,$$

$$\mu = \langle X \rangle = \int_{s=t}^{s=T} f(s)\,ds,$$

$$\sigma^2 = \langle X^2 \rangle - \mu^2 = \int_{s=t}^{s=T} g^2(s)\,ds.$$

The whole analysis was made simple because $f(t)$ and $g(t)$ depended only on time (and not on the stochastic value itself, X).

Solution to Exercise 9

The process for S is obtained by applying straight differential (form) analysis plus the special "Ito's lemma term" to get:

$$dS = \left[\frac{\partial S(x)}{\partial x}\,dx + \frac{1}{2}\frac{\partial^2 S(x)}{\partial x^2}\sigma^2(x,t)\,dt\right]_{x=\ln S}$$

$$= S\left(\mu(\ln S, t) + \frac{\sigma(\ln S, t)^2}{2}\right)dt + S\sigma(\ln S, t)\,dz(t).$$

Solution to Exercise 10

Aside on Vector Analysis   . If two vectors in three dimensional space are given by $\bar{a} = (a_1, a_2, a_3)$ and $\bar{b} = (b_1, b_2, b_3)$, then the measure of whether they are perpendicular is the *dot product* defined as

$$\bar{a} \cdot \bar{b} = a_1 b_1 + a_2 b_2 + a_3 b_3$$

and they *are* perpendicular if this has the value zero. Further two vectors, $\bar{a}$ and $\bar{b}$, may be combined to form a third vector, which is perpendicular to them both, called the *cross-product,* that is

$$\bar{a} \times \bar{b} = (a_2 b_3 - a_3 b_2, a_3 b_1 - a_1 b_3, a_1 b_2 - a_2 b_1).$$

The two processes require three securities,

$$dS_1 = S_1\mu_1\,dt + S_1(\sigma_{11}\,dz_1(t) + \sigma_{12}\,dz_2(t)),$$

$$dS_2 = S_2\mu_2\,dt + S_2(\sigma_{21}\,dz_1(t) + \sigma_{22}\,dz_2(t)),$$

$$dS_3 = S_3\mu_3\,dt + S_3(\sigma_{31}\,dz_1(t) + \sigma_{32}\,dz_2(t)),$$

and we may then construct a portfolio

$$\Pi = \Delta_1 S_1 + \Delta_2 S_2 + \Delta_3 S_3$$

and this results in a process for the portfolio

$$
\begin{aligned}
d\Pi &= \Delta_1 dS_1 + \Delta_2 dS_2 + \Delta_3 dS_3 \\
&= \Delta_1(S_1\mu_1\,dt + S_1(\sigma_{11}dz_1(t) + \sigma_{12}dz_2(t))) \\
&\quad + \Delta_2(S_2\mu_2\,dt + S_2(\sigma_{21}dz_1(t) + \sigma_{22}dz_2(t))) \\
&\quad + \Delta_3(S_3\mu_3\,dt + S_3(\sigma_{31}dz_1(t) + \sigma_{32}dz_2(t))).
\end{aligned}
$$

To make this process nonstochastic we require choosing the deltas so that the coefficients of the two drivers are *both* zero independently:

$$\Delta_1 S_1\sigma_{11} + \Delta_2 S_2\sigma_{21} + \Delta_3 S_3\sigma_{31} = 0,$$

$$\Delta_1 S_1\sigma_{12} + \Delta_2 S_2\sigma_{22} + \Delta_3 S_3\sigma_{32} = 0.$$

The solution to exercise 10 is that the "vector" $(\Delta_1 S_1, \Delta_2 S_2, \Delta_3 S_3)$ must be perpendicular to the "plane" defined by the two "vectors" $(\sigma_{11}, \sigma_{21}, \sigma_{31})$ and $(\sigma_{12}, \sigma_{22}, \sigma_{32})$. This is solved by

$$
\begin{pmatrix} S_1\Delta_1 \\ S_2\Delta_2 \\ S_3\Delta_3 \end{pmatrix} = \alpha \begin{pmatrix} \sigma_{21}\sigma_{32} - \sigma_{22}\sigma_{31} \\ \sigma_{12}\sigma_{31} - \sigma_{11}\sigma_{32} \\ \sigma_{11}\sigma_{22} - \sigma_{12}\sigma_{21} \end{pmatrix}
$$

for any value α. Now, if the portfolio has no stochastic behavior, it means that the drift term becomes equal to the riskless yield rate on the portfolio minus stock borrow fees and so

$$\mu_1 S_1\Delta_1 + \mu_2 S_2\Delta_2 + \mu_3 S_3\Delta_3 = r\Pi - r_1 S_1\Delta_1 - r_2 S_2\Delta_2 - r_3 S_3\Delta_3$$

$$\Rightarrow (\mu_1 - r + r_1)S_1\Delta_1 + (\mu_2 - r + r_2)S_2\Delta_2 + (\mu_3 - r + r_3)S_3\Delta_3 = 0.$$

In turn, this formula means that the vector

$$((\mu_1 - r - r_1), (\mu_2 - r - r_2), (\mu_3 - r - r_3))$$

is perpendicular to the vector $(\Delta_1 S_1, \Delta_2 S_2, \Delta_3 S_3)$ and so must lie in the plane defined by the two vectors above; $(\sigma_{11}, \sigma_{21}, \sigma_{31})$ and $(\sigma_{12}, \sigma_{22}, \sigma_{32})$. Thus

$$(\mu_1 - r + r_1) = \lambda_1 \sigma_{11} + \lambda_2 \sigma_{12},$$

$$(\mu_2 - r + r_2) = \lambda_1 \sigma_{21} + \lambda_2 \sigma_{22},$$

$$(\mu_3 - r + r_3) = \lambda_1 \sigma_{31} + \lambda_2 \sigma_{32},$$

which is the result that there is a universal market risk premium associated with each driver and the securities' individual "volatility" to that driver. The result generalizes to an unlimited number of drivers.

Solution to Exercise 11

1. $E(S_T > K) = \int_{z=z_K}^{\infty} P(z)dz = N(z_k)$
   where

$$z_K = \frac{\ln\left(\frac{K}{S}\right) - MT}{\Sigma\sqrt{T}},$$

$$M = \frac{1}{T} \int_{t=0}^{t=T} \left(\mu(t) - \frac{\sigma(t)^2}{2}\right) dt,$$

$$\Sigma^2 = \frac{1}{T} \int_{t=0}^{t=T} \sigma(t)^2 \, dt,$$

$$N(x) = \frac{1}{\sqrt{2\pi}} \int_{z=-\infty}^{z=x} \exp\left(-\frac{z^2}{2}\right) dz.$$

2. This probability is the term that multiplies the strike in the expectation value formula.

Solution to Exercise 12

Forward Kolmogorov equation as

$$\left(-\frac{\partial}{\partial T} + \frac{S_T^2 \sigma_0^2}{2} \frac{\partial^2}{\partial S_T^2} - S_T(\mu_0 - 2\sigma_0^2)\frac{\partial}{\partial S_T} + (\sigma_0^2 - \mu_0)\right) f(S_T, T) = 0.$$

The solutions are

$$\exp(-(\mu_0 - \sigma_0^2)(T - t)),$$
$$S_T \exp(-(2\mu_0 - 3\sigma_0^2)(T - t)),$$
$$S_T^2 \exp(-(3\mu_0 - 6\sigma_0^2)(T - t)).$$

Solution to Exercise 13

1. Backward Kolmogorov equation written as

$$\left(\frac{\partial}{\partial t} + \frac{S_t^2 \sigma_0^2}{2} \frac{\partial^2}{\partial S_t^2} + S_t \mu_0 \frac{\partial}{\partial S_t} \right) f(S_t, t) = 0,$$

and has solutions expressed as

$$1,$$
$$S_t \exp(\mu_0(T - t)),$$
$$S_t^2 \exp\left(2 \left(\mu_0 + \frac{\sigma_0^2}{2} \right)(T - t) \right)$$

2. The backward Kolmogorov equation solutions describe the expected future path of stock prices that start at S_t at time t: the mean of these paths for $T > t$ is

$$\langle S_T \rangle = S_t \exp(\mu_0(T - t)),$$

and the standard deviation is

$$\sqrt{\frac{\langle S_T^2 \rangle}{\langle S_T \rangle^2}} = \exp\left(\frac{\sigma_0^2}{2}(T - t) \right).$$

The forward Kolmogorov equation describes the paths that arrive at the point S_T at time T as

$$\exp(-(\mu_0 - \sigma_0^2)(T - t)),$$
$$S_T \exp(-(2\mu_0 - 3\sigma_0^2)(T - t)),$$
$$S_T^2 \exp(-(3\mu_0 - 6\sigma_0^2)(T - t)).$$

The first term implies the distribution is not correctly normalized because the condition all *past* paths that arrive at S_T is not a complete set on the probability measure (which is not surprising as many paths miss S_T). The average value of these paths at some time t for $t < T$, is then

$$\langle S_t \rangle = \frac{S_T \exp(-(2\mu_0 - 3\sigma_0^2)(T - t))}{\exp(-(\mu_0 - \sigma_0^2)(T - t))} = S_T \exp(-(\mu_0 - 2\sigma_0^2)(T - t)).$$

The standard deviation is

$$\sqrt{\frac{\langle S_t^2 \rangle}{\langle S_t \rangle^2}} = \sqrt{\left(\frac{S_T^2 \exp(-(3\mu_0 - 6\sigma_0^2)(T - t))}{\exp(-(\mu_0 - \sigma_0^2)(T - t))} \right) \frac{\exp(-(\mu_0 - \sigma_0^2)(T - t))}{S_T \exp(-(2\mu_0 - 3\sigma_0^2)(T - t))}}$$

$$= \exp\left(\frac{1}{2}\sigma_0^2(T - t) \right).$$

Solution to Exercise 14

Given a boundary function $S_t = S'(t)$, solve the backward Kolmogorov differential equation with the boundary condition that $f(S_t, t)|_{S_t = S'(t)} = t$ as

$$\left(\frac{\partial}{\partial t} + \frac{S_t^2 \sigma_0^2}{2} \frac{\partial^2}{\partial S_t^2} + S_t \mu_0 \frac{\partial}{\partial S_t} \right) f(S_t, t) = 0.$$

Solution to Exercise 15

Given a boundary function $S_T = S'(T)$, solve the forward Kolmogorov differential equation with the boundary condition that $f(S, T)|_{S = S'(T)} = T$ as

$$\left(-\frac{\partial}{\partial T} + \frac{S_T^2 \sigma_0^2}{2} \frac{\partial^2}{\partial S_T^2} - S_T(\mu_0 - 2\sigma_0^2)\frac{\partial}{\partial S_T} + (\sigma_0^2 - \mu_0) \right) f(S_T, T) = 0.$$

Solution to Exercise 16

None! The stock-price process, and premium process may be written as

$$dS = S\mu \, dt + S\sigma \, dz_1(t),$$

$$d\lambda = \lambda\mu_\lambda \, dt + \lambda\sigma_\lambda dz_2(t).$$

The development of the pricing equation requires two instruments to hedge with but the result (refer to Exercise 11's solution) is that

$$(\mu - r + r_1) = \lambda\sigma,$$

$$(\mu_\lambda - r + r_\lambda) = \lambda_2\sigma_\lambda,$$

$$(\mu_3 - r + r_3) = \lambda\sigma_{option-s} + \lambda_2\sigma_{option-\lambda}.$$

Because the call option has no part of its payout dependent on risk premium, the risk premium, curvature terms may be dropped from the pricing equation for call options and so, effectively,

$$\sigma_{option-\lambda} = 0.$$

The result would be different for derivatives that have a payout dependent on the risk premium at expiration.

Another way to see this is that the Black-Scholes pricing formulas for calls and puts apply to all worlds with various values for the risk premium. If the risk premium varies along a path it does not affect the option price nor the hedging strategy and further, if we then average over many paths there is still no effect on the option price.

Solution to Exercise 17

The variance of the hedging strategy exactly cancels the variance of the expectation value. Because there is only one unique function that has the variance of the expectation value at maturity, that is, square of the payout, the expectation value of this function (the square of the payout) must be the variance of the expectation value itself. So

$$\langle \text{var(hedging strategy)} \rangle = E(payoff^2, \lambda) - E(payoff, \lambda)^2.$$

Using a final payout $f(S,T) = \max((S-K)^2, 0)$, and a process $dS = S\mu\, dt + S\sigma\, dz_1(t)$ solve:

$$\left(\frac{\partial}{\partial t} + \frac{S^2\sigma^2}{2}\frac{\partial^2}{\partial S^2} + S\mu\frac{\partial}{\partial S} - r_\$ \right) f(S,t) = 0$$

where

$$t < T \text{ and } \mu = r_\$ - r_1 + \lambda\sigma.$$

Then

$$E(payoff^2, \lambda; t < T) = f(S,t).$$

Solution to Exercise 18

The Risk Premium derivation of section 4.1 made no mention of which security was the derivative and which was the underlying, only that the market had one driver and some volatility and drift for each security. This means that the final result holds for underlyings or derivatives symmetrically. Section 6.3 Black-Scholes Equation: Relation to Risk Premium Definition shows that the Black-Scholes equation is just the imposition of the constraint of the definition of the market risk premium applying to the derivative as well as the underlying. This whole derivation is then completely symmetric and can be interpreted the other way around by switching the derivative and the underlying, and their corresponding borrow rates.

Solution to Exercise 19

First, there is only one second-order derivative. So we look for the variable change to make this into a straight second-order derivative. The easiest way to do this, in fact, is to note that a stochastic process corresponds to this equation as a backward Kolmogorov equation with a stochastic term of the form

$$dS = \ldots + S\sigma \, dz(t)$$

and this can be rewritten more simply (ignoring drift terms and Ito's lemma terms) by integrating

$$\frac{dS}{S\sigma}$$

to get a variable

$$x = \int \frac{1}{S\sigma} \, dS \sim \frac{1}{\sigma} \ln S,$$

$$S = \exp(x\sigma).$$

And then use this variable instead of S, by substituting into the differential equation,

$$\frac{\partial}{\partial x} = \frac{\partial S}{\partial x} \frac{\partial}{\partial S} = \sigma \exp(\sigma x) \frac{\partial}{\partial S}$$

$$\Rightarrow \frac{\partial}{\partial S} = \frac{\exp(-\sigma x)}{\sigma} \frac{\partial}{\partial x}$$

$$\Rightarrow \frac{\partial^2}{\partial S^2} = \frac{\exp(-2\sigma x)}{\sigma^2} \frac{\partial^2}{\partial x^2} - \frac{\exp(-2\sigma x)}{\sigma} \frac{\partial}{\partial x},$$

which we can then use in the equation operator, expressed as

$$\left(\frac{\partial}{\partial t} + \frac{S^2\sigma^2}{2}\frac{\partial^2}{\partial S^2} + S\mu\frac{\partial}{\partial S} - r_\$\right)f(S,t) = 0$$

$$\Rightarrow \left(\frac{\partial}{\partial t} + \frac{1}{2}\frac{\partial^2}{\partial x^2} + \frac{\mu}{\sigma}\frac{\partial}{\partial x} - r_\$\right)F(x,t) = 0$$

where

$$F(x,t) = f(S(x),t).$$

Solution to Exercise 20

The Fourier transform of a put option payout is as follows:

$$x = \ln(S/K) + \left(r_\$ - r_{st} - \frac{\sigma^2}{2}\right)(T-t),$$

$$
\begin{aligned}
f_T(k) &= \frac{1}{\sqrt{2\pi}} \int_{x=-\infty}^{x=\infty} f_T(x)\exp(-ikx)\,dx \\
&= \frac{K}{\sqrt{2\pi}} \int_{x=-\infty}^{x=\infty} \max(1-e^x,0)\exp(-ikx)\,dx \\
&= \frac{K}{\sqrt{2\pi}} \int_{x=-\infty}^{x=0} (e^x-1)\exp(-ikx)\,dx \\
&= \frac{K}{\sqrt{2\pi}} \left[\frac{\exp((1-ik)x)}{1-ik} - \frac{\exp(-ikx)}{-ik}\right]_{x=-\infty}^{0} \\
&= \frac{K}{\sqrt{2\pi}} \left(\frac{1}{1-ik} - \frac{i}{k}\right).
\end{aligned}
$$

The last step can be understood only by using the Lebesgue integral (see section 2.3 Functional Analysis and Fourier Transforms), where we have to introduce a *regularizing* function inside the integral to make the integrand go to zero for large values of x. This function might be

$$g(x) = \exp\left(\frac{1}{(x+a)(x-a)}\right)$$

and then take the limit $a \to \infty$ after integrating. The result is the last line above.

Solution to Exercise 21

The Fourier transform of the Black-Scholes put option formula immediately follows

$$P(k, T - t; K) = \frac{K}{\sqrt{2\pi}} \left(\frac{1}{1 - ik} - \frac{i}{k} \right) \exp \left(-\frac{\sigma^2 k^2 (T - t)}{2} \right).$$

Solution to Exercise 22

First the European call option formula satisfies

$$\left(\frac{\partial}{\partial t} + \frac{S^2 \sigma^2}{2} \frac{\partial^2}{\partial S^2} + S(r_\$ - r_S) \frac{\partial}{\partial S} - r_\$ \right) C_E(S, t; T, \sigma^2, r_\$, r_S) = 0,$$

and an American option, $C_A(S,t; T, \sigma^2, r_\$, r_S)$, satisfies the same equation below the stock price exercise boundary $S = S_*(t)$. Thus writing

$$C_A(S, t; T, \sigma^2, r_\$, r_S) = C_E(S, t; T(S, t), \sigma^2, r_\$, r_S).$$

Then

$$\left(\frac{\partial T}{\partial t} \frac{\partial C_E}{\partial T} + \frac{S^2 \sigma^2}{2} \left\{ 2 \frac{\partial^2 C_E}{\partial S \partial T} \frac{\partial T}{\partial S} + \frac{\partial^2 C_E}{\partial T^2} \left(\frac{\partial T}{\partial S} \right)^2 + \frac{\partial C_E}{\partial T} \frac{\partial^2 T}{\partial S^2} \right\} \right.$$

$$\left. + S(r_\$ - r_S) \left\{ \frac{\partial C_E}{\partial T} \frac{\partial T}{\partial S} \right\} \right) = 0,$$

which implies

$$T(S, T) = T,$$

$$T(S_*, t) = t,$$

$$T(0, t) = T,$$

$$\left. \begin{array}{c} S < S_*(t) \\ t < T \end{array} \right\} : \frac{\partial T}{\partial t} + \frac{S^2 \sigma^2}{2} \frac{\partial^2 T}{\partial S^2} + S(r_\$ - r_S + \mu) \frac{\partial T}{\partial S} + \omega \left(\frac{\partial T}{\partial S} \right)^2 = 0,$$

where

$$\mu = \mu(S, t) = \sigma^2 S \frac{\partial}{\partial S} \ln \left(\frac{\partial C_E}{\partial T} \right),$$

$$\omega = \omega(S, t) = \frac{S^2 \sigma^2}{2} \frac{\partial}{\partial T} \ln \left(\frac{\partial C_E}{\partial T} \right).$$

Solution to Exercise 23

The Bond formula is given by

$$B_T(t) = \sum_{i=1}^{N} c_i \exp\left[-\int_{s=t}^{s=t_i} \left(r(s) + \frac{\phi(s)}{1 - R(s)}\right) ds\right]$$

$$+ \int_{s=t}^{s=T} \frac{\phi(s)R(s)}{1 - R(s)} \exp\left[-\int_{u=t}^{u=s} \left(r(u) + \frac{\phi(u)}{1 - R(u)}\right) du\right] ds.$$

The recovery term may then be broken down by interval

$$\text{Recovery}_T(t) = \int_{s=t}^{s=T} \frac{\phi(s)R(s)}{1 - R(s)} \exp\left[-\int_{u=t}^{u=s} \left(r(u) + \frac{\phi(u)}{1 - R(u)}\right) du\right] ds$$

$$= \sum_{i=1}^{i=N} \int_{s=t_{i-1}}^{s=t_i} \frac{\phi(s)R(s)}{1 - R(s)} \exp\left[-\int_{u=t}^{u=s} \left(r(u) + \frac{\phi(u)}{1 - R(u)}\right) du\right] ds,$$

where the sum is taken over all the sections of the union of curve dates for all the curves, and thus all the curves are flat over any interval.

$$\text{Recovery}_T(t) = \sum_{i=1}^{i=N} \int_{s=t_{i-1}}^{s=t_i} \frac{\phi_i R_i}{1 - R_i} \exp\left[-\int_{u=t}^{u=s} \left(r(u) + \frac{\phi(u)}{1 - R(u)}\right) du\right] ds$$

$$= \sum_{i=1}^{i=N} \int_{s=t_{i-1}}^{s=t_i} \frac{\phi_i R_i}{1 - R_i} \exp\left[-\int_{u=t_{i-1}}^{u=s} \left(r_i + \frac{\phi_i}{1 - R_i}\right) du\right.$$

$$\left. - \int_{u=t}^{u=t_{i-1}} \left(r(u) + \frac{\phi(u)}{1 - R(u)}\right) du\right] ds$$

$$= \sum_{i=1}^{i=N} \exp\left[-\int_{u=t}^{u=t_{i-1}} \left(r(u) + \frac{\phi(u)}{1 - R(u)}\right) du\right]$$

$$\frac{\phi_i R_i}{1 - R_i} \int_{s=t_{i-1}}^{s=t_i} \exp\left[-\left(r_i + \frac{\phi_i}{1 - R_i}\right)(s - t_{i-1})\right] ds$$

and so

$$\text{Recovery}_T(t) = \sum_{i=1}^{i=N} \exp\left[-\int_{u=t}^{u=t_{i-1}} \left(r(u) + \frac{\phi(u)}{1-R(u)}\right) du\right]$$

$$\times \frac{\phi_i R_i}{1-R_i} \left[\frac{e^{-\left(r_i + \frac{\phi_i}{1-R_i}\right)(s-t_{i-1})}}{-\left(r_i + \frac{\phi_i}{1-R_i}\right)}\right]_{s=t_{i-1}}^{s=t_i}$$

and we can perform this integral easily,

$$\text{Recovery}_T(t) = \sum_{i=1}^{i=N} \left\{ \frac{\phi_i R_i}{1-R_i} \left(\frac{1 - e^{-\left(r_i + \frac{\phi_i}{1-R_i}\right)(t_i - t_{i-1})}}{\left(r_i + \frac{\phi_i}{1-R_i}\right)}\right) \sum_{j=1}^{j=i-1} \right.$$

$$\left. \exp\left[-\left(r_j + \frac{\phi_j}{1-R_j}\right)(t_j - t_{j-1})\right]\right\}.$$

The second term is a discounting term and the first term is thus the value of recovery over an interval where all the curves are flat; that is, the expression is a present value on the first date of the interval. This is then discounted to today and these contributions to the total recovery value are then added up over all intervals.

The conclusion is that piecewise flat forward curves make the analysis of recovery very tractable.

Solution to Exercise 24

First the variable change needs to be implemented and we know by symmetry that all the αs and ps are the same, that is

$$\langle p_1 \rangle^{(N)} = [1 + (N-1)\alpha]p\Delta t,$$

$$\langle p_1 p_2 \rangle^{(N)} = [2\alpha + (N-2)\alpha^2]p\Delta t,$$

$$\langle p_1 p_2 \ldots p_m \rangle^{(N)} = [m\alpha^{m-1} + (N-m)\alpha^m]p\Delta t.$$

Then we immediately find the variable change values,

$$\phi = [1 + (N-1)\alpha]p, \quad \rho = \frac{2\alpha + (N-2)\alpha^2}{1 + (N-1)\alpha},$$

$$\Rightarrow \alpha = \frac{\sqrt{\rho^2(N-1)^2 + 4(1-\rho)} - [2 - \rho(N-1)]}{2(N-2)}, \quad p = \frac{\phi}{(1 + (N-1)\alpha)}$$

with the exception of the case for $N = 2$ for which

$$\alpha = \frac{\rho}{(2 - \rho)}.$$

Making use of the identities

$$(\beta + \alpha)^N = \sum_{m=0}^{m=N} \frac{N!}{(N - m)!m!} \alpha^m \beta^{N-m},$$

$$N(\beta + \alpha)^{N-1} = N\beta^{N-1} + \sum_{m=1}^{m=N-1} \frac{N!}{(N - m - 1)!m!} \alpha^m \beta^{N-m-1},$$

$$N(\beta + \alpha)^{N-1} = \sum_{m=1}^{m=N} \frac{N!}{(N - m)!(m - 1)!} \alpha^{m-1} \beta^{N-m}$$

and given that

$$\langle s_1 s_2 \ldots s_K \rangle^{(N)} - \exp\left(-\int F_K^{(N)} \, dt\right),$$

$$F_K^{(N)} - \sum_{m=1}^{m=K} (-)^{m+1} \sum_{perms} \langle p_1 p_2 \ldots p_m \rangle^{(N)},$$

$$= \sum_{m=1}^{m=K} (-)^{m+1} \frac{K!}{(K - m)!m!} [m\alpha^{m-1}p + (N - m)\alpha^m p]$$

we find that the K identical bond correlated-default survival-probability is given by

$$F_K^{(N)} = \sum_{m=1}^{m=K} (-)^{m-1} \frac{K!}{(K - m)!m - 1!} \alpha^{m-1}p - N\sum_{m=1}^{m=K} (-)^m \frac{K!}{(K - m)!m!} \alpha^m p$$

$$- \alpha \sum_{m=1}^{m=K} (-)^{m-1} \frac{K!}{(K - m)!m - 1!} \alpha^{m-1}p$$

$$= K(1 - \alpha)^{K-1} - N[(1 - \alpha)^K - 1] - \alpha K(1 - \alpha)^{K-1}$$

$$\Rightarrow F_K^{(N)} = (N - (N - K)(1 - \alpha)^K)p = \frac{(N - (N - K)(1 - \alpha)^K)}{(1 + (N - 1)\alpha)} \frac{\phi}{(1 - R)}.$$

Consider that the K-bond correlated survival probability, $\langle s_1 s_2 \ldots s_K \rangle^{(N)}$, *is* dependent on N in general. This is not surprising given the observation that while two bonds may have no intrinsic interaction (i.e., $\alpha = 0$), because both are correlated to a third bond, this gives them a correlation. Also consider that for absolutely no correlation, $\rho = 0$, we get the K-bond conditional survival probability as expected

$$\langle s_1 s_2 \ldots s_K \rangle^{(N)} = \exp\left(-\int \frac{K\phi}{(1-R)}\, dt\right)$$

and for perfect correlation, $\rho = 1$, the conditional probability acts as a single bond default probability, expressed as

$$\langle s_1 s_2 \ldots s_K \rangle^{(N)} = \exp\left(-\int \frac{\phi}{(1-R)}\, dt\right) = \langle s_1 \rangle^{(N)}.$$

Central Limit Theorem-Plausibility Argument

I f a path begins at a point x_0 then the initial distribution over x is the Dirac delta distribution, which means merely that the path has probability 1 of being located exactly at the starting point, expressed as

$$p(x; t = 0) - \delta(x - x_0).$$

Then use a *square shock,* meaning that the distribution is convoluted with a square of width 2α and height $1/(2\alpha)$,

$$G(x_1, x_2; \Delta t) = \begin{cases} \frac{1}{2\alpha} & -\alpha \le |x_2 - x_1| \le \alpha; \\ 0 & otherwise \end{cases}$$

and the new distribution is then a square. Then convolute again and the new distribution is a triangle,

$$G(x_1, x_3; 2\Delta t) = \int_{x_1=-\infty}^{x_1=+\infty} G(x_1, x_2; \Delta t) G(x_2, x_3; \Delta t)\, dx_2$$

$$= \begin{cases} \frac{1}{2\alpha} - \frac{|x_2 - x_1|}{4\alpha^2} & -2\alpha \le (x_2 - x_1) \le 2\alpha; \\ 0 & otherwise \end{cases}$$

and so on. Now, in more general terms, let the shock be some function $f(x)$, not necessarily a square. Fourier transform the function to get $f(p)$, and expand it as follows:

$$f(p) = a_0 \exp(-a_2(p - p_0)^2 + a_3(p - p_0)^3 + \cdots)$$

about its maximum, at p_0. This expansion is meaningful as long as the derivative of $f(p)$ is zero and continuous at p_0 and the second derivative of

$f(p)$ is continuous and negative (i.e., $a_2 > 0$). Then take the N-th power of this, to get the Fourier transform of the operator that gives us the N-step distribution:

$$FT[G(x_1, x_2; t_2 - t_1)] = [f(p)]^N$$
$$= a_0^N \exp(-a_2 N(p - p_0)^2 + a_3 N(p - p_0)^3 + \cdots).$$

Now the argument is that because the first term looks like the Fourier transform of a Gaussian of standard deviation

$$\Sigma = \sigma \sqrt{N} = \sqrt{2a_2 N},$$

and for a particular standard deviation, the first term has size

$$|p - p_0|^2 \Sigma^2$$

and we require the first term be small, ε^2 say, then

$$|p - p_0| < \frac{\varepsilon}{\Sigma},$$

which means that the region of relevant p (i.e., Fourier modes) gets smaller and smaller corresponding to a larger and larger standard deviation, but also corresponding to a fixed dimension relative to the Gaussian. The second term has size

$$a_3 N |p - p_0|^3 < \frac{N \varepsilon^3 a_3}{\Sigma^3} = \left(\frac{\varepsilon^2 a_3}{\sigma^3}\right) \frac{\varepsilon}{\sqrt{N}},$$

which implies that N can always be made large enough to suppress the second term relative to the first, when we are looking at scales bigger than 1 standard deviation. The solution looks like the Fourier transform of a Gaussian of standard deviation $\sqrt{2a_2 N}$, in the large N limit.

Imposing the constraint that the expansion be well defined implies that the second derivative of the Fourier transform near the maximum is defined, which is the same as saying (choosing $p_0 = 0$), as follows:

$$\frac{d}{dp}G(p)\Big|_{p=0} = \int_{p=-\infty}^{p=+\infty} \frac{d}{dp}G(p)\delta(p)\,dp = 0,$$

$$-\int_{p=-\infty}^{p=+\infty} \frac{d^2}{dp^2}G(p)\delta(p)\,dp = -const$$

$$\Rightarrow$$

$$\int_{x=-\infty}^{x=\infty} xG(x;t)\,dx - 0,$$

$$\int_{x=-\infty}^{x=\infty} x^2G(x;t)\,dx = const.$$

That is, the variance is well defined. The choice $p_0 = 0$ corresponded to a particular choice of x variables (i.e., nondrifting ones) and so, in general, the central limit theorem works for distributions with well-defined norm, mean, and standard deviation:

$$\int_{x=-\infty}^{x=\infty} G(x;t)\,dx = 1,$$

$$\int_{x=-\infty}^{x=\infty} xG(x;t)\,dx = \bar{x},$$

$$\int_{x=-\infty}^{x=\infty} (x-\bar{x})^2G(x;t)\,dx = const.$$

Notes Relevant to Finance

1. The state density has to be 1, otherwise it shows up inside the expansion, and then the result without the density function corresponds to a different distribution. This different limiting distribution is none other than a variable change of a normal distribution.
2. If the shocks are Gaussians, we may note that the sum of Gaussians is a Gaussian. It is the only distribution, with defined variance, for which this is true.
3. If the variance of the single shock is

$$\int_{x=-\infty}^{x=\infty} x^2G(x;\Delta t)\,dx$$

(i.e., zero mean) then the variance of the limiting Gaussian is

$$N \int_{x=-\infty}^{x=\infty} x^2 G(x; \Delta t)\, dx,$$

that is, the standard deviation is preserved in this sense.

4. If there is a nonzero drift for each single step

$$\int_{x=-\infty}^{x=\infty} x G(x; \Delta t)\, dx,$$

then $|p_0| > 0$ and this results in the Fourier modes being multiplied by $e^{i p \bar{x}}$. But this in turn implies that the function mean is shifted. If the single shock has a nonzero mean

$$\int_{x=-\infty}^{x=\infty} x G(x; \Delta t)\, dx,$$

then the limiting Gaussian has mean

$$N \int_{x=-\infty}^{x=\infty} x G(x; \Delta t)\, dx.$$

The standard deviation of the Gaussian is still $\sqrt{N}$ times the shock standard deviation.

Note that the square shock example given at the beginning of the section requires that

$$\alpha = \sigma \sqrt{3 \Delta t}$$

to ensure the square distribution

$$G(x_1, x_2; \Delta t) = \begin{cases} \frac{1}{2\alpha} & -\alpha \le |x_2 - x_1| \le \alpha; \\ 0 & \textit{otherwise} \end{cases}$$

has standard deviation $\sigma \sqrt{\Delta t}$, and that it therefore tends to a Gaussian of standard deviation $\sigma \sqrt{t}$ as the number of steps gets very large.

5. A class of distributions (often called *fat-tailed*) fall off as a power of x, $\frac{1}{x^\lambda}$ for $0 \le \lambda \le 3$. The standard deviation is not defined (it is not finite). They do not produce exactly the same functional forms over the entire range of the state-variable because the convolution is repeated, but their tails do look similar as they propagate.

Levy distributions are a special class of these power law tail distributions because they do produce convolutions that have the same functional form as the individual shocks. They form the basis for a lot of so-called fat-tail modeling in finance. The Cauchy-Levy distribution is a special case of this class of distributions.

Solving for the Green's Function of the Black-Scholes Equation

W e want to solve

$$\frac{\partial G(x - x_0, t - t_0)}{\partial t} + \frac{\sigma^2}{2} \frac{\partial^2 G(x - x_0, t - t_0)}{\partial x^2} = \delta(t - t_0)\delta(x - x_0).$$

We can set the reference coordinates to zero because they can be added later by subtracting from the solution's arguments. This is because the differential equation does not contain any explicit functions of the coordinates—that is, it is translation invarian. Thus

$$\frac{\partial G(x, t)}{\partial t} + \frac{\sigma^2}{2} \frac{\partial^2 G(x, t)}{\partial x^2} = \delta(t)\delta(x).$$

Take the 2-D Fourier transform of the equation, that is, $(x, t) \mapsto (k, \omega)$,

$$i\omega G(k, \omega) - \frac{\sigma^2 k^2}{2} G(k, \omega) = \frac{1}{2\pi}$$

$$\Rightarrow \quad G(k, \omega) = \frac{1}{2\pi \left(i\omega - \frac{\sigma^2 k^2}{2} \right)}.$$

Then

$$G(x, t) = \frac{1}{2\pi} \iint_{k, \omega} G(k, \omega) e^{ikx + i\omega t} \, dk \, d\omega$$

$$= \frac{1}{2\pi} \iint_{k, \omega} \frac{e^{ikx + i\omega t}}{2\pi \left(i\omega - \frac{\sigma^2 k^2}{2} \right)} \, dk \, d\omega.$$

The residue theorem from complex analysis means that the result is only nonzero for closing the integral over ω in the lower half of the complex ω-plane. This happens only if $t < 0$. Thus

$$G(x,t) = \frac{-i}{2\pi}H(-t)\int_{k=-\infty}^{k=+\infty} e^{ikx+\varpi(k)t}\,dk,$$

$$\varpi(k) = \frac{\sigma^2 k^2}{2}.$$

By completing the square (recalling $t < 0$), we can factor out a function of x and t times an integral,

$$z = \sigma k\sqrt{-t} - \frac{ix}{\sigma\sqrt{-t}}$$

$$\Rightarrow G(x,t) = H(-t)\frac{1}{2\pi\sigma\sqrt{-t}}\exp\left[\frac{x^2}{2\sigma^2 t}\right]\int_{z=-\infty}^{z=+\infty}\exp\left(-\frac{z^2}{2}\right)dz$$

$$= H(-t)\frac{1}{\sqrt{2\pi}\sigma\sqrt{-t}}\exp\left[\frac{x^2}{2\sigma^2 t}\right].$$

Finally substitute $t \to t - T$, and $x \to x - x_T$ to get

$$G(x,t;x_T,T) = H(T-t)\frac{1}{\sqrt{2\pi\sigma^2(T-t)}}\exp\left[-\frac{(x-x_T)^2}{2\sigma^2(T-t)}\right].$$

Expanding the von Neumann Stability Mode for the Discretized Black-Scholes Equation

From section 9.3.2 Gaussian Kurtosis (and Skew $= 0$), Faster Convergence, the von Neumann stability mode $C_{i,j} = \zeta^{\frac{T}{\Delta t}} e^{ikx}$ is

$$C_{i-1,j} = \exp(-r_\$ \Delta t)(p_{j+1} C_{i-1,j+1} + p_j C_{i-1,j} + p_{j-1} C_{i-1,j-1}),$$

$$p_{j-1} = \left(-\alpha e^{\Delta x} + e^{\frac{\Delta x}{2}} \beta\right) / \det, \quad p_{j+1} = \left(-\alpha e^{-\Delta x} + e^{\frac{-\Delta x}{2}} \beta\right) / \det,$$

$$p_j = 1 - p_{j+1} - p_{j-1},$$

$$\alpha = 2 \sinh(\Delta x)(e^{r_{short}\Delta t} - 1), \quad \beta = 2 \sinh\left(\frac{\Delta x}{2}\right)(e^{(\sigma^2 + 2r_{short})\Delta t} - 1),$$

$$\det = 2 \sinh(2\Delta x) - 4 \sinh(\Delta x);$$

$$\Rightarrow \zeta = \exp(-r_\$ \Delta t)(p_{j+1} e^{ik\Delta x} + p_j + p_{j-1} e^{-ik\Delta x}) = \exp(-r_\$ \Delta t)\zeta_0.$$

Then using only the up and down probabilities,

$$\zeta_0 = p_{j+1} e^{ik\Delta x} + p_j + p_{j-1} e^{-ik\Delta x} = 1 + p_{j+1}(e^{ik\Delta x} - 1)$$
$$+ p_{j-1}(e^{-ik\Delta x} - 1)$$
$$= 1 + (p_{j+1} + p_{j-1})(\cos(k\Delta x) - 1) + i(p_{j+1} - p_{j-1})\sin(k\Delta x)$$
$$p_{j+1} + p_{j-1} = [-2 \sinh(2\Delta x)(e^{r_{short}\Delta t} - 1) + 2 \sinh(\Delta x)$$
$$\times (e^{(\sigma^2 + 2r_{short})\Delta t} - 1)]\frac{1}{\det}$$

$$p_{j+1} - p_{j-1} = \left[4\sinh^2(\Delta x)(e^{r_{short}\Delta t} - 1) - 4\sinh^2\left(\frac{\Delta x}{2}\right) \right.$$
$$\left. \times \; (e^{(\sigma^2 + 2r_{short})\Delta t} - 1) \right] \frac{1}{\det}.$$

Next, expanding in powers of Δx,

$$\det = 2\Delta x^3 + \frac{1}{2}\Delta x^5 + \cdots, \quad p = \frac{\sigma^2 \Delta t}{2\Delta x^2},$$

$$p_{j+1} + p_{j-1} = \frac{\Delta t \sigma^2}{\Delta x^2} - \left(r_{short} + \frac{\sigma^2}{12} \right) \Delta t$$
$$+ \frac{\Delta t^2}{\Delta x^2} \left(r_{short}^2 + 2r_{short}\sigma^2 + \frac{\sigma^4}{2} \right) + \cdots,$$

$$p_{j+1} - p_{j-1} = \frac{\Delta t}{2\Delta x}(2r_{short} - \sigma^2) + \frac{\Delta x \Delta t}{3}\left(r_{short} + \frac{\sigma^2}{4} \right)$$
$$- \frac{\Delta t^2 \sigma^2}{\Delta x}\left(r_{short} + \frac{\sigma^2}{4} \right) + \cdots;$$

$$(p_{j+1} + p_{j-1})(\cos(k\Delta x) - 1)$$
$$= (p_{j+1} + p_{j-1})\frac{-k^2\Delta x^2}{2}\left(1 - \frac{k^2\Delta x^2}{12} + \cdots \right)$$
$$= \left[-\frac{k^2\sigma^2\Delta t}{2} + \frac{k^4\sigma^4\Delta t^2}{48p} + \frac{k^2\sigma^2}{4p}\left(r_{short} + \frac{\sigma^2}{12} \right)\Delta t^2 \right.$$
$$\left. - \frac{k^2\Delta t^2}{2}\left(r_{short}^2 + 2r_{short}\sigma^2 + \frac{\sigma^4}{2} \right) + \cdots \right];$$

$$i(p_{j+1} - p_{j-1})\sin(k\Delta x)$$
$$= i(p_{j+1} - p_{j-1})k\Delta x\left(1 - \frac{k^2\Delta x^2}{6} + \cdots \right)$$
$$= i\left[k\Delta t(r_{short} - \sigma^2/2) - \frac{\sigma^2 k^3 \Delta t^2}{12p}(r_{short} - \sigma^2/2) \right.$$
$$\left. + k\Delta t^2\sigma^2\left(\frac{1}{6p} - 1 \right)\left(r_{short} + \frac{\sigma^2}{4} \right) + \cdots \right]$$

which, in turn, leads to

$$
\begin{aligned}
\zeta_0 = 1 &+ \left(-\frac{\sigma^2 k^2}{2} + ik \left(r_{short} - \frac{\sigma^2}{2} \right) \right) \Delta t \\
&+ \frac{1}{2} \left(-\frac{\sigma^2 k^2}{2} + ik \left(r_{short} - \frac{\sigma^2}{2} \right) \right)^2 \Delta t^2 \\
&+ \frac{1}{2} \left(\frac{1}{6p} - 1 \right) \left\{ \left(\frac{\sigma^2 k^2}{2} \right)^2 - i\sigma^2 k^3 \left(r_{short} - \frac{\sigma^2}{2} \right) \right. \\
&+ \left. \sigma^2 k^2 \left(3r_{short} + \frac{\sigma^2}{4} \right) + i\sigma^2 k \left(2r_{short} + \frac{\sigma^2}{2} \right) \right\} \Delta t^2 + O(\Delta t^3).
\end{aligned}
$$

Multiple Bond Survival Probabilities Given Correlated Default Probability Rates

L et the survival probability of bond 1 be s_1 and the default probability be $p_1 \Delta t$ over a time period Δt, then

$$s_1 = 1 - p_1 \Delta t.$$

If we introduce the idea of an expectation probability for default, then

$$\langle p_1 \rangle^{(1)} = p_1 \Delta t,$$

$$\langle s_1 \rangle^{(1)} = s_1 = 1 - p_1 \Delta t$$

where the superscript denotes the fact that the probability is in a system with only one variable (bond).

We introduce correlation for a pair of bonds 1 and 2, by making the double default pobability different to the product of the uncorrelated probabilities by introducing a "mixing parameter" α_{12},

$$\langle p_1 p_2 \rangle^{(2)} = (p_1 \Delta t)(p_2 \Delta t) + \alpha_{12}(p_1 \Delta t s_2 + s_1 p_2 \Delta t),$$

$$\langle p_1 \rangle^{(2)} = p_1 \Delta t + \alpha_{12} s_1 p_2 \Delta t.$$

This means that the probability of a double default is not the same as the product of the probabilities of the separate defaults in a world where the bonds are not correlated. We have related a world with correlation to one without correlation. α_{12} is itself a kind of probability. The probability that the event counted by single default in the uncorrelated system actually results in a double default in the correlated system. We have, in fact, performed a variable change from an uncorrelated set of variables to a correlated set.

It follows from the definition of correlation that

$$\rho_{12}^{(2)} = \frac{\langle p_1 p_2 \rangle^{(2)} - \langle p_1 \rangle^{(2)} \langle p_2 \rangle^{(2)}}{\sqrt{\langle p_1 \rangle^{(2)} \langle p_2 \rangle^{(2)} \langle s_1 \rangle^{(2)} \langle s_2 \rangle^{(2)}}}.$$

Expanding to lowest order in Δt (and setting recovery rates to zero for simplicity for the moment) we find that

$$\langle p_1 p_2 \rangle^{(2)} \approx \alpha_{12}(p_1 + p_2)\Delta t,$$

$$\langle p_1 \rangle^{(2)} \approx (p_1 + \alpha_{12}p_2)\Delta t = \phi_1 \Delta t,$$

$$\langle p_1 \rangle^{(2)} \approx (p_2 + \alpha_{12}p_1)\Delta t = \phi_2 \Delta t,$$

$$\rho_{12}^{(2)} \approx \frac{\alpha_{12}(p_1 + p_2)}{\sqrt{\phi_1 \phi_2}}.$$

What is happening here is that a variable change from $\{\phi_1, \phi_2, \rho_{12}^{(2)}\}$ to $\{p_1, p_2, \alpha_{12}\}$ is occurring, expressed as

$$\phi_1 = p_1 + \alpha_{12}p_2,$$

$$\phi_2 = p_2 + \alpha_{12}p_1,$$

$$\rho_{12}^{(2)} = \frac{\alpha_{12}(p_1 + p_2)}{\sqrt{(p_1 + \alpha_{12}p_2)(p_2 + \alpha_{12}p_1)}}$$

$$\Rightarrow$$

$$\alpha_{12} = \frac{\sqrt{\phi_1 \phi_2}\rho_{12}^{(2)}}{(\phi_1 + \phi_2 - \rho_{12}^{(2)}\sqrt{\phi_1 \phi_2})},$$

$$p_1 = \frac{\phi_1 - \alpha_{12}\phi_2}{(1 - \alpha_{12}^2)},$$

$$p_2 = \frac{\phi_2 - \alpha_{12}\phi_1}{(1 - \alpha_{12}^2)}.$$

This enables us to construct the survival probabilities for a two-bond system (denoted with the superscript) to first order:

$$\langle s_1 \rangle^{(2)} = 1 - \phi_1 \Delta t,$$

$$\langle s_2 \rangle^{(2)} = 1 - \phi_2 \Delta t,$$

$$\langle s_1 s_2 \rangle^{(2)} = 1 - [p_1 + p_2]\Delta t = 1 - [\phi_1 + \phi_2 - \rho_{12}^{(2)}\sqrt{\phi_1 \phi_2}]\Delta t$$

and thus expressions for correlated survival probabilities over finite times (t to T say),

$$\langle s_1 \rangle^{(2)} = \exp\left(-\int_{s=t}^{s=T} \phi_1(s)ds\right), \quad \langle s_2 \rangle^{(2)} = \exp\left(-\int_{s=t}^{s=T} \phi_2(s)ds\right);$$

$$\langle s_1 s_2 \rangle^{(2)} = \exp\left(-\int_{s=t}^{s=T} [\phi_2(s) + \phi_1(s) - \rho_{12}^{(2)}(s)\sqrt{\phi_1(s)\phi_2(s)}]ds\right).$$

Moving to a three-bond system: note that even if $\alpha_{12} = 0$ the variables 1 and 2 *are* correlated. Counting all the outcomes using

$$1 = p_1 p_2 p_3 + s_1 p_2 p_3 + p_1 s_2 p_3 + p_1 p_2 s_3$$
$$+ p_1 s_2 s_3 + s_1 p_2 s_3 + s_1 s_2 p_3 + s_1 s_2 s_3,$$

we find that

$$\langle p_1 \rangle^{(3)} = [p_1 + \alpha_{12} s_1 p_2 s_3 + \alpha_{13} s_1 s_2 p_3]\Delta t$$
$$+ (\alpha_{12} + \alpha_{13} - \alpha_{12}\alpha_{13})s_1 p_2 p_3 \Delta t^2,$$
$$\langle p_1 p_2 \rangle^{(3)} = [\alpha_{12} p_1 s_2 s_3 + \alpha_{12} s_1 p_2 s_3 + \alpha_{23}\alpha_{13} s_1 s_2 p_3]\Delta t$$
$$+ [p_1 p_2 + (\alpha_{12} + \alpha_{13} - \alpha_{12}\alpha_{13})s_1 p_2 p_3$$
$$+ (\alpha_{12} + \alpha_{23} - \alpha_{12}\alpha_{23})p_1 s_2 p_3]\Delta t^2.$$

Thus

$$\rho_{12}^{(3)}\sqrt{\langle p_1 \rangle^{(3)}\langle p_2 \rangle^{(3)}\langle s_1 \rangle^{(3)}\langle s_2 \rangle^{(3)}} = \langle p_1 p_2 \rangle_{(3)} - \langle p_1 \rangle^{(3)}\langle p_2 \rangle^{(3)}$$

and if we set $\alpha_{12} = 0$ then we find, even to first order in default, that

$$\rho_{12}^{(3)}\sqrt{\langle p_1 \rangle^{(3)}\langle p_2 \rangle^{(3)}\langle s_1 \rangle^{(3)}\langle s_2 \rangle^{(3)}} \neq 0.$$

The conclusion is that we have to be very careful relating the αs to the correlation parameters.

To first order in default, that is, Δt, we find that

$$\langle p_1 \rangle^{(3)} = (p_1 + \alpha_{12}p_2 + \alpha_{13}p_3)\Delta t = \phi_1 \Delta t,$$

$$\langle p_2 \rangle^{(3)} = (p_2 + \alpha_{12}p_1 + \alpha_{23}p_3)\Delta t = \phi_2 \Delta t,$$

$$\langle p_3 \rangle^{(3)} = (p_3 + \alpha_{13}p_1 + \alpha_{23}p_2)\Delta t = \phi_3 \Delta t,$$

$$\langle p_1 p_2 \rangle^{(3)} = (\alpha_{12}(p_1 + p_2) + \alpha_{23}\alpha_{13}p_3)\Delta t,$$

$$\langle p_1 p_2 p_3 \rangle^{(3)} = (\alpha_{12}\alpha_{13}p_1 + \alpha_{12}\alpha_{23}p_2 + \alpha_{13}\alpha_{23}p_3)\Delta t,$$

$$\rho_{12}^{(3)} = \frac{(\alpha_{12}(p_1 + p_2) + \alpha_{23}\alpha_{13}p_3)}{\sqrt{\phi_1 \phi_2}}.$$

Again, the reality is that we are implementing a variable change from correlated variables $\{\phi_1, \phi_2, \phi_3, \rho_{12}^{(3)}, \rho_{13}^{(3)}, \rho_{23}^{(3)}\}$ to uncorrelated variables $\{p_1, p_2, p_3, \alpha_{12}^{(3)}, \alpha_{13}^{(3)}, \alpha_{23}^{(3)}\}$, expressed as

$$\phi_1 = p_1 + \alpha_{12}p_2 + \alpha_{13}p_3,$$

$$\phi_2 = \alpha_{12}p_1 + p_2 + \alpha_{23}p_3,$$

$$\phi_3 = \alpha_{13}p_1 + \alpha_{23}p_2 + p_3;$$

$$\rho_{12}^{(3)}\sqrt{\phi_1 \phi_2} = \alpha_{12}(p_1 + p_2) + \alpha_{13}\alpha_{23}p_3,$$

$$\rho_{23}^{(3)}\sqrt{\phi_2 \phi_3} = \alpha_{23}(p_2 + p_3) + \alpha_{12}\alpha_{13}p_1,$$

$$\rho_{13}^{(3)}\sqrt{\phi_1 \phi_3} = \alpha_{13}(p_1 + p_3) + \alpha_{23}\alpha_{12}p_2.$$

Reversing this we find that

$$p_1 = \frac{\phi_1(1 - \alpha_{23}^2) - \phi_2(\alpha_{12} - \alpha_{23}\alpha_{13}) - \phi_3(\alpha_{13} - \alpha_{23}\alpha_{12})}{(1 - \alpha_{13}^2 - \alpha_{12}^2 - \alpha_{23}^2 + 2\alpha_{23}\alpha_{12}\alpha_{13})},$$

$$p_2 = \frac{-\phi_1(\alpha_{12} - \alpha_{13}\alpha_{23}) + \phi_2(1 - \alpha_{13}^2) - \phi_3(\alpha_{23} - \alpha_{13}\alpha_{12})}{(1 - \alpha_{13}^2 - \alpha_{12}^2 - \alpha_{23}^2 + 2\alpha_{23}\alpha_{12}\alpha_{13})},$$

$$p_3 = \frac{-\phi_2(\alpha_{23} - \alpha_{12}\alpha_{13}) - \phi_1(\alpha_{13} - \alpha_{12}\alpha_{23}) + \phi_3(1 - \alpha_{12}^2)}{(1 - \alpha_{13}^2 - \alpha_{12}^2 - \alpha_{23}^2 + 2\alpha_{23}\alpha_{12}\alpha_{13})}.$$

The expression for the α_{ij}'s is complicated, particularly by being nonlinear, but it is solvable up to $(N - 1)$ parameters for each α_{ij}, namely, $N(N - 1)(N - 1)/2$ parameters in total, which correspond to angles on a sphere in p-space because each of the $N(N-1)/2$ correlation-defining equations describes an elliptical surface in that space.

This leads to the evaluation of the default and survival probabilities as functions of the input variables:

$$\langle s_1 \rangle^{(3)} = 1 - \phi_1 \Delta t,$$

$$\langle s_2 \rangle^{(3)} = 1 - \phi_2 \Delta t,$$

$$\langle s_3 \rangle^{(3)} = 1 - \phi_3 \Delta t,$$

$$\langle s_1 s_2 \rangle^{(3)} = 1 - [\phi_1 + \phi_2 - \sqrt{\phi_1 \phi_2} \rho_{12}^{(3)}] \Delta t,$$

$$\langle s_1 s_2 s_3 \rangle^{(3)} = 1 - [\phi_1 + \phi_2 + \phi_3 - \sqrt{\phi_1 \phi_2} \rho_{12}^{(3)}$$
$$- \sqrt{\phi_2 \phi_3} \rho_{23}^{(3)} - \sqrt{\phi_1 \phi_3} \rho_{13}^{(3)} + \cdots] \Delta t.$$

The extra terms are not easily writable in terms of the input variables, but their value is well defined:

$$\phi_1 = p_1 + \alpha_{12} p_2 + \alpha_{13} p_3,$$

$$\phi_2 = p_2 + \alpha_{12} p_1 + \alpha_{23} p_3,$$

$$\phi_3 = p_3 + \alpha_{13} p_1 + \alpha_{23} p_2;$$

$$\rho_{12} \sqrt{\phi_1 \phi_2} = \alpha_{12}(p_1 + p_2) + \alpha_{23} \alpha_{13} p_3,$$

$$\rho_{23} \sqrt{\phi_2 \phi_3} = \alpha_{23}(p_2 + p_3) + \alpha_{12} \alpha_{13} p_1,$$

$$\rho_{13} \sqrt{\phi_1 \phi_3} = \alpha_{13}(p_1 + p_3) + \alpha_{12} \alpha_{23} p_2;$$

$$\langle p_1 \rangle^{(3)} = (p_1 + \alpha_{12} p_2 + \alpha_{13} p_3) \Delta t = \phi_1 \Delta t,$$

$$\langle p_1 p_2 \rangle^{(3)} = (\alpha_{12}(p_1 + p_2) + \alpha_{23} \alpha_{13} p_3) \Delta t = \rho_{12} \sqrt{\phi_1 \phi_2} \Delta t,$$

$$\langle p_1 p_2 p_3 \rangle^{(3)} = (\alpha_{12} \alpha_{13} p_1 + \alpha_{12} \alpha_{23} p_2 + \alpha_{13} \alpha_{23} p_3) \Delta t.$$

Once more the first six equations determine the $\{p_i\}$ and $\{\alpha_{ij}\}$ variables as functions of the input variables $\{\phi_i\}$ and $\{\rho_{ij}\}$ and thus the term $\langle p_1 p_2 p_3 \rangle^{(3)}$ is expressible as a function of the input variables.

We can now easily extend the analysis to N bonds as

$$\phi_i = p_i + \sum_{\substack{j=1 \\ j \neq i}}^{j=N} \alpha_{ij} p_j,$$

$$\rho_{ij}^{(N)} = \frac{1}{\sqrt{\phi_i \phi_j}} \left(\alpha_{ij}(p_i + p_j) + \sum_{\substack{k=1 \\ k \neq i,j}}^{k=N} \alpha_{ik}\alpha_{jk}p_k \right)$$

which when inverted gives the $\{p_i\}$ and $\{\alpha_{ij}\}$ variables as functions of the input variables $\{\phi_i\}$ and $\{\rho_{ij}\}$ and then

$$\langle p_i \rangle^{(N)} = p_i \Delta t + \sum_{\substack{j=1 \\ j \neq i}}^{j=N} \alpha_{ij}p_j \Delta t,$$

$$\langle p_i p_j \rangle^{(N)} \approx \left[\alpha_{ij}(p_i + p_j) + \sum_{\substack{k=1 \\ k \neq i,j}}^{k=N} \alpha_{ik}\alpha_{jk}p_k \right] \Delta t,$$

$$\langle p_1 p_2 p_3 \ldots p_m \rangle^{(N)} \approx [\alpha_{12}\alpha_{13}\ldots\alpha_{1m}]p_1 \Delta t + [\alpha_{12}\alpha_{23}\alpha_{24}\ldots\alpha_{2m}]p_2 \Delta t$$

$$+ \cdots + [\alpha_{1m}\alpha_{1m}\ldots\alpha_{(m-1)m}]p_m \Delta t$$

$$+ \sum_{k \notin \{a,b,c\ldots,m\}} [\alpha_{ka}\alpha_{kh}\alpha_{kc}\ldots\alpha_{km}]p_k \Delta t,$$

which enables us to value the correlated survival probabilities as follows:

$$\langle s_1 s_2 s_3 \ldots s_m \rangle^{(N)} = \exp\left(-\int_{s=t}^{s=T} F_m^N(s)ds \right),$$

$$F_m^N(t) = \left[\langle p_i \rangle^{(N)} - \langle p_i p_j \rangle^{(N)} + \langle p_i p_j p_k \rangle^{(N)} \right.$$

$$\left. + \cdots + \langle p_1 p_2 \ldots p_m \rangle^{(N)} \right]_{sum\ over\ perms\{i,j,k\ldots,m\}}.$$

Note

$$F_N^N(t) = \sum_{j=1}^{j=N} p_j.$$

So the algorithm for calculating correlated survival probabilities is: Use the input variables (curves) $\phi_i(t)$, $\rho_{ij}^{(N)}(t)$ to find the uncorrelated variables $p_i(t)$, $\alpha_{ij}^{(N)}(t)$; then use these variables to find the F curve using the formula above.

References

Abramowitz, Milton, and Irene A. Stegun. 1972. *Handbook of Mathematical Functions.* Washington, DC: Dover.

Barone-Adesi, G., and R. E. Whaley. 1987. *Efficient analytic approximation of American optionvalues.* Journal of Finance 42: 301-20.

Courant, Richard, and David Hilbert. 1989. *Methods of mathematical physics*, Vols. I and II. New York, NY: J. Wiley & Sons.

Cheyette, O. "Term Structure Dynamics and Mortgage Valuation." *Journal of Fixed Income* 2 (1992): 28–41.

Duffie, Darrell. 2001. *Dynamic asset pricing theory.* Princeton, NJ: Princeton University Press.

Heath, D., R. Jarrow, and A. Morton. "Bond Pricing and the Term Structure of Interest Rates: A New Methodology for Contingent Claim Valuation." *Econometrica* 60 (1992): 77–105.

Hull, John C. 2005. *Options, futures and other derivatives*, 6th edn. Upper Saddle River, NJ: Prentice Hall.

Ingersoll, Jonathan E. 1987. *Theory of financial decision making.* Lanham, MD: Rowman & Littlefield.

Li, D. "On default correlation: A Copula function approach." *Journal of Fixed Income* 9 (2000): 43–45.

Index

A

Abramowitz, Milton, 178
American boundary (free boundary),
    97–98
American exercise, 97–100
American option, 158
American-style exercise, 117
American-style option, 4
Analytic bond call option, 74–75
Analytic formula. See European option
Analytic zero-coupon bond valuation,
    73–74
Arbitrage freedom, 77
Arbitrage-free value, 113
ATM. See At-the-money
At-the-market CDS, 111
At-the-money (ATM) index options,
    122
Average annualized volatility, 41

B

Backward Kolmogorov equation,
    43–46
    solution, 52, 142, 153
    type, 51–53
    writing, 153
Barone-Adesi, G., 178
Black-Scholes call option formula,
    Fourier transform, 84
Black-Scholes derivation, 49–50

Black-Scholes equation, 51–53
    analytical solution, 81–84
    application. See Currency options
    derivation, 46–48
    discussion, 81
    financial interpretation, 48
    Green's function, solving, 167–168
    index borrow inputs, 122
    numeric solution, 84–104
    regaining, 61
    relationship. See Risk-neutral pricing;
        Risk premium
    rewriting, 84–85
    solutions, 97, 117
        alternative, 52
    transformation, 60–61
    understanding, 51
    usage, 48. See also Martingale;
        Options
    values, 99
    volatility inputs, 121
    von Neumann stability mode,
        expansion. See Discretized
        Black-Scholes equation
Black-Scholes formula, 5
    assumptions, 6
    modifications, 7
    rewriting, 60
    usage, 123–124
Black-Scholes option pricing formulae,
    51–52

Printed and bound by CPI Group (UK) Ltd, Croydon, CR0 4YY

27/10/2024

14580318-0001